# Applied ethology 2018
## Ethology for health and welfare

# ISAE2018

Proceedings of the
## 52nd Congress of the International Society for Applied Ethology

July 30th to August 3rd, 2018
Charlottetown, Prince Edward Island, Canada

# ETHOLOGY FOR HEALTH AND WELFARE

edited by:

Michael Cockram
Tarjei Tennessen
Luis Bate
Renée Bergeron
Sylvie Cloutier
Andrew Fisher
Maria Hötzel

OASES

Online Academic Submission and Evaluation System

EAN: 9789086863228
e-EAN: 9789086868704
ISBN: 978-90-8686-322-8
e-ISBN: 978-90-8686-870-4
DOI: 10.3920/978-90-8686-870-4

**First published, 2018**

© **Wageningen Academic Publishers**
**The Netherlands, 2018**

Wageningen Academic
P u b l i s h e r s

# Welcome to the 52<sup>nd</sup> Congress of ISAE

We welcome you to Charlottetown, Prince Edward Island, Canada, and hope that you will enjoy the conference and the Island. It has been an honour and a pleasure to make the arrangements for this conference. We have prepared an interesting program based on your submissions and a series of events that will provide entertainment and an opportunity to experience the unique local culture.

The main theme of the Congress, "Ethology for Health and Welfare," was chosen to reflect the prominence that applied ethology has in the field of animal welfare and to encourage the development of applied ethology in studies to promote animal health. The location of this year's Congress within the Atlantic Veterinary College at the University of Prince Edward Island has provided the focus on veterinary aspects of ethology and welfare. Canada is blessed with abundant and diverse wildlife that has been managed by humans throughout the country's history. These interactions are integral to Canadian culture and traditions. We have taken the opportunity to highlight how ethological studies have contributed to the management and welfare of wild animals.

As scientific studies have advanced our understanding of applied ethology, a major role of the ISAE is to apply our science to real-world issues that affect the management of animals. Our session on putting ethology into practice is designed to stimulate discussion of best practices and challenges in achieving this goal. Applied ethology continues to develop and expand, and we have showcased recent developments in play behaviour and other key topics. The Wood-Gush Lecture will provide the stimulation that comes from developments and approaches used in allied disciplines and the challenges that can arise from differing thinking on some potential implications of our increased understanding of how animals perceive and experience their world.

We are grateful for the sponsorship that we have received, for the help from colleagues at the University of Prince Edward Island and Dalhousie University, and to the University of Prince Edward Island for agreeing to host this year's Congress. We thank you for travelling to Atlantic Canada and hope that your experiences here will be memorable.

*Michael Cockram* and *Tarjei Tennessen*

# Acknowledgements

## Local organising committee

Michael Cockram, Tarjei Tennessen, Luis Bate, Sam Buchanan, Linda Constable, Jeffrey Davidson, Marc Doiron, Robert Gilmour, Miriam Gordon, Myrtle Jenkins-Smith, Greg Keefe, Cathy Ryan, and Jo-Ann Thomsen

## Scientific committee

Michael Cockram, Tarjei Tennessen, Luis Bate, Renée Bergeron, Sylvie Cloutier, Andrew Fisher, and Maria Hötzel

## ISAE Ethics Committee

Anna Olsson, Alexandra Whittaker, Francois Martin, Franck Peron, Francesco De Giorgio, Cecilie Mejdell, and Elize van Vollenhoven

## Student helpers

Megan Jewell, Alysha Trenholm, Jenna MacKay, Tiffany Belbeck, Amy Harrington, and Sabrina Nicholson

## ISAE 2018 conference design, abstract book cover, and UPEI maps

UPEI Marketing and Communications

# Reviewers

Siobhan Abeyesinghe
Michael Appleby
Ngaio Beausoleil
Bjarne Braastad
Clover Bench
Jennifer Brown
Stephanie Buijs
Joao Costa
Ingrid de Jong
Nicolas Devillers
Trevor DeVries
Catherine Dwyer
Hans Erhard
Bjorn Forkman
Derek Haley
Alison Hanlon
Alexandra Harlander
Moira Harris
Marie Haskell
Camie Heleski
Paul Hemsworth
Mette Herskin
Michelle Hunniford
Margit Bak Jensen
Linda Keeling
Meagan King
Ute Knierim
Peter Krawczel
Jan Langbein
Alistair Lawrence
Donald Lay Jr.
Caroline Lee
Lena Lidfors
Jeremy Marchant-Forde
Danila Marini

Lindsay Matthews
Rebecca Meagher
Suzanne Millman
Daniel Mills
Lene Munksgaard
Ruth Newberry
Christine Nicol
Lee Niel
Birte Nielsen
Cheryl O'Connor
Anna Olsson
Kathryn Proudfoot
Jean-Loup Rault
Bas Rodenburg
Yolande Seddon
Marek Spinka
Hans Spoolder
Mhairi Sutherland
Stephanie Torrey
Michael Toscano
Cassandra Tucker
Patricia Turner
Simon Turner
Frank Tuyttens
Anna Valros
Nienke van Staaveren
Eberhard von Borell
Marina von Keyserlingk
Susanne Waiblinger
Daniel Weary
Laura Webb
Francoise Wemelsfelder
Tina Widowski
Christoph Winckler

**Sponsors**

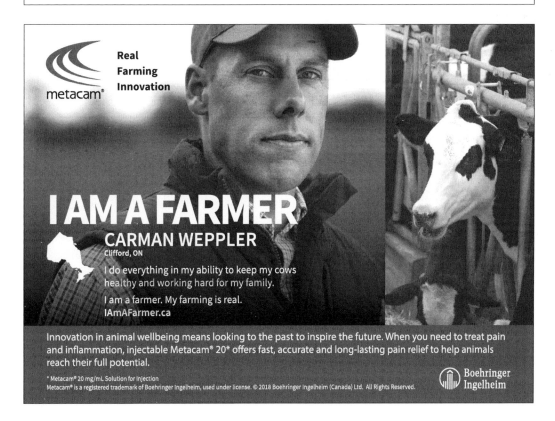

# Campus maps

## CAMPUS BUILDINGS

1  Chi-Wan Young Sports Centre (YSC)
2  Central Utility Building (CUB)
3  Health Sciences Building (HSB)
4  W.A. Murphy Student Centre (MSC)
5  SDU Main Building (SDMB)
6  Steel Building (SB)
7  Dalton Hall (DH)
9  Memorial Hall (MH)
10 Cass Science Hall (CSH)
11 Kelley Memorial Building (KMB)

12 Don and Marion McDougall Hall (MCDH)
13 Duffy Science Centre (DSC)
14 Chaplaincy Centre (CC)
15 Robertson Library (RL)
16 Daycare Building (DCB)
17 K.C. Irving Chemistry Centre (ICC)
18 Wanda Wyatt Dining Hall (WDH)
19 Bill and Denise Andrew Hall (AH) (Residence)
20 Bernardine Hall (BEH) (Residence)
24 Blanchard Hall (BLH) (Residence)

27 Atlantic Veterinary College (AVC)
28 Regis and Joan Duffy Research Centre (DRC)
30 Faculty of Sustainable Design Engineering
33 Artificial Turf Field Announcers' Building
34 Clubhouse
35 Alumni Canada Games Place Storage Building
36 Alumni Canada Games Place Announcers' Building
37 Alumni Canada Games Place VIP Building
38 Development and Alumni Engagement Building

  BUS SHELTER
★ EMERGENCY CALL STATION

## PARKING

A  General & Overnight During Winter Months
B  General & Designated
C  Designated
D  General
E  General
F  Designated
G  Designated
VTH Teaching Hospital Clients
♿ Accessible
RO Residence Only
VP Visitor Metered (UPEI parking pass not required)
RP Reserved
SR Shipping and Receiving
RC Research Centre

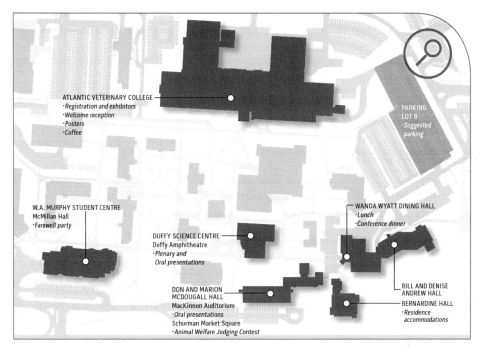

Locaton of buildings used on UPEI campus

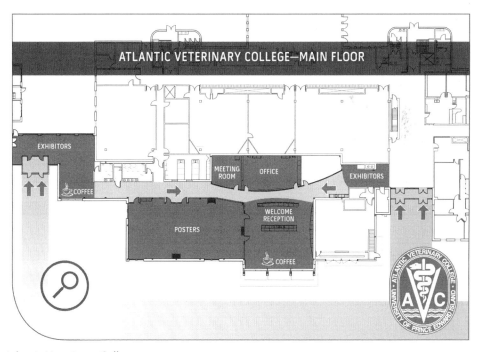

Atlantic Veterinary College

# General information

## Venues

The Congress will be held at the University of Prince Edward Island, 550 University Avenue, Charlottetown, Prince Edward Island, C1A 4P3, Canada.

## Official language

English is the official language of the ISAE 2018 Congress.

## Registration and information desk

The registration desk will be located in the main lobby of the Atlantic Veterinary College before the welcome reception on Monday, July 30th, and from Tuesday, July 31st, to Friday August 3rd, in Room 279N, Atlantic Veterinary College. The opening hours of the information desk will be indicated on-site. The Atlantic Veterinary College can be entered via the main entrance and the entrance to the North Annex.

## Name badges

Name badges are required to allow admittance to the Congress sessions, coffee breaks, lunches, welcome reception, animal welfare judging contest, and farewell party. Badges will be issued at registration.

## Posters

Posters will be shown in Rooms 286N and 287N of the Atlantic Veterinary College.

## Exhibition area

Sponsors and other exhibitors will be located in the foyer areas of the Atlantic Veterinary College.

## Internet access codes

Instructions on how to access the internet will be provided at registration.

## Coffee breaks

Tea, coffee, and refreshments will be served in the lobby of the Atlantic Veterinary College with access via the main entrance and the entrance to the North Annex, in Rooms 285N and 2301N, either side of the poster presentations, and near the registration desk.

## Lunches

Lunches will be held in the Wanda Wyatt Dining Hall. Meal tickets for lunch will be provided in the registration package.

## Parking

Delegates who have vehicles may park in one of the non-gated parking lots bearing the letters A, B, D, or E. It will be easier to enter the campus via the Belvedere Avenue entrance rather than the University Avenue entrance.

# Social program

### Welcome reception, Monday July 30th, 18.00 to 21.00 h

The welcome reception will be held in the lobby of the Atlantic Veterinary College with access via the main entrance and the entrance to the North Annex, in Rooms 285N and 2301N, either side of the poster presentations, and near the registration desk. ISAE student representatives will provide a welcome to other graduate students and suggest networking tips between 19.00 and 20.30 h. Tickets for the welcome reception are included in the Congress registration fee, and entry is open to all registered delegates and registered accompanying persons.

### Wine and cheese poster session, Tuesday July 31st, 17.15 to 19.00 h

A poster session will be held in Rooms 286N and 287N of the Atlantic Veterinary College with food and drink served in adjacent areas. Participation is open to all registered delegates.

### Animal Welfare Judging Contest, Tuesday July 31st, 19.00 to 21.00 h

The animal welfare judging contest will take place in the Alex H. MacKinnon Auditorium and Schurman Market Square, both located in Don and Marion McDougall Hall. Participation is open to all registered delegates.

### Excursions, Wednesday August 1st, 14.00 to 20.00 h

Conference excursions will take place during the afternoon and evening. Coaches will leave from UPEI to various locations on PEI, and all tours will conclude with a lobster supper/vegan stir fry in the evening at North Rustico before returning to UPEI at 20.00 h.

### Faculty-Student Lunchtime Session, Thursday August 2nd, 13.00 to 14.30 h

*Eating with Ethologists*: On Thursday, the ISAE student representatives will match students and faculty with similar interests for a networking lunch in the Wanda Wyatt Dining Hall.

### Conference dinner, Thursday August 2nd, 18.30 to 23.59 h

The conference dinner will be held in the Wanda Wyatt Dining Hall. Tickets must be purchased in advance via the on-line booking system.

## Student Lunchtime Session, Friday August 3<sup>rd</sup>, 13.00 to 14.30 h

On Friday, the ISAE student representatives will arrange tables to facilitate discussions on specific topics during lunch in the Wanda Wyatt Dining Hall. This will be combined with the Global Development Lunch.

## Farewell party, Friday August 3<sup>rd</sup>, 18.30 to 23. 00 h

The farewell party will be held in McMillan Hall, W.A. Murphy Student Centre.
Tickets are included in the Congress registration fee, and entry is open to all registered delegates and registered accompanying persons.

# Scientific program

## Monday July 30<sup>th</sup>

14.00–21.00 **Registration**, Atlantic Veterinary College

14.00–21.00 **Poster placement**, Atlantic Veterinary College

18.00–21.00 **Welcome reception**, Atlantic Veterinary College

## Tuesday July 31<sup>st</sup>

8.00 Poster placement, Atlantic Veterinary College . . . . . . . . . . . . . . . . . . . . . . . . . . . . . . . . . . . . . . . . .

9.00 **Opening Ceremony**, Duffy Science Centre Amphitheatre

9.30 **Wood-Gush Memorial Lecture,** Duffy Science Centre Amphitheatre
The other-minds problem in other species. *Stevan Harnad*

11.00 Break and posters, Atlantic Veterinary College . . . . . . . . . . . . . . . . . . . . . . . . . . . . . . . . . . . . . . . .

| Session 1 | Session 2 |
|---|---|
| **Cognition and emotions** | **Equine behaviour and welfare** |
| Duffy Science Centre Amphitheatre | Alex H. MacKinnon Auditorium, Don and Marion McDougall Hall |

| | | |
|---|---|---|
| 11.45 | Cognitive evaluation of predictability and controllability and implications for animal welfare. *Caroline Lee* | An acoustic indicator of positive emotions in horses? *Mathilde Stomp* |
| 12.15 | Understanding the bidirectional relationship of emotional and cognitive systems to measure affective state in the animal. *Sebastian McBride* | Sleep vs. no sleep; equine nocturnal behavior and its implications for equine welfare. *Linda Greening* |
| 12.30 | Evaluation of sheep anticipatory behaviour and lateralisation by means of fNIRS technique. *Matteo Chincarini* | Shelter-seeking behaviour in domestic donkeys and horses in a temperate climate. *Britta Osthaus* |
| 12.45 | Assessing differential attention to simultaneous negative and positive video stimuli in sheep: moving toward attention bias. *Camille Raoult* | Social referencing in the domestic horse. *Christian Nawroth* |

13.00 Lunch, Wanda Wyatt Dining Hall . . . . . . . . . . . . . . . . . . . . . . . . . . . . . . . . . . . . . . . . . . . . . . . . . . . . .

# Tuesday July 31st

| | Session 3<br>**Putting ethology into practice**<br>Duffy Science Centre Amphitheatre | |
|---|---|---|

| 14.30 | Crossing the divide between academic research and practical application of ethology on commercial livestock and poultry farms. *Temple Grandin* |

| | Session 3 (continued) | Session 4<br>**Laboratory animal: behaviour and welfare**<br>Alex H. MacKinnon Auditorium, Don and<br>Marion McDougall Hall |
|---|---|---|
| 15.15 | Integrating ethology into actual practice on Canadian farms. *Jacqueline Wepruk* | Female mice can distinguish between conspecifics raised with enrichment and those raised in standard 'shoebox' cages. *Aimée Marie Adcock* |
| 15.30 | Balancing outcome-based measures and resource-based measures in broiler welfare policies: the NGO perspective. *Monica List* | Enriched mice are nice: long-term effects of environmental enrichment on agonism in female C57BL/6s, DBA/2s, and BALB/cs. *Emma Nip* |

| 15.45 | **Break and posters, Atlantic Veterinary College** ........................................... |
|---|---|

| 16.30 | A global perspective on environmental enrichment for pigs: how far have we come? *Heleen Van De Weerd* | Early pup mortality in laboratory mouse breeding: presence of older pups critically affects perinatal survival. *Sophie Brajon* |
|---|---|---|
| 16.45 | Establishing science-based standards for the care and welfare of dogs in US commercial breeding facilities. *Candace Croney* | The best things come in threes: Assessing recommendations to minimize male mouse aggression. *Brianna Gaskill* |
| 17.00 | A rotation designed to teach final-year veterinary students about dairy cattle welfare. *Todd Duffield* | Behavioural evidence of curiosity in Zebrafish. *Becca Franks* |

| 17.15 | **Wine & Cheese Poster session, Atlantic Veterinary College** .................................. |
|---|---|

| 19.00–21.00 | **Animal Welfare Judging Contest**<br>Alex H. MacKinnon Auditorium and Schurman Market Square, Don and Marion McDougall Hall |
|---|---|

# Wednesday August 1st

| | Session 5 **Cattle behaviour and welfare** Duffy Science Centre Amphitheatre | Session 6 **Companion animal behaviour and welfare** Alex H. MacKinnon Auditorium, Don and Marion McDougall Hall |
|---|---|---|
| 9.00 | Seek and hide: Understanding pre-partum behavior of cattle by use of inter-species comparison. *Maria Vilain Rørvang* | Turning off dogs' brains: fear of noises affects problem solving behavior and locomotion in standardized cognitive tests. *Karen Overall* |
| 9.30 | Effect of parity on dairy cows and heifers' preferences for calving location. *Erika Edwards* | Improving canine welfare in commercial breeding (CB) operations: evaluating rehoming candidates. *Judith Stella* |
| 9.45 | Use of Qualitative Behaviour Assessment to evaluate styles of maternal protective behaviour in Girolando Cows. *Maria Camila Ceballos* | The predictive validity of an observational shelter dog behaviour assessment. *Conor Goold* |
| 10.00 | The effect of concrete feedpad feeding periods on oestrous behaviour in dairy cows in a pasture-based farming system. *Andrew Fisher* | Refining on-site canine welfare assessment: evaluating the reliability of Field Instantaneous Dog Observation (FIDO) scoring. *Lynda Mugenda* |
| 10.15 | Effects of three types of bedding surface on dairy cattle behavior, preference, and hygiene. *Karin Schütz* | Dimensions of personality in working dogs: an analysis of collected data on future guide and assistance dogs over 37 years. *Nicolas Dollion* |
| 10.30 | The impact of housing tie-stall-housed dairy cows in deep-bedded loose-pens during the dry period on lying time and postures. *Elise Shepley* | The effects of aversive- and reward-based training methods on companion dog welfare. *Anna Olsson* |
| 10.45 | Cross-contextual vocalisation rates and heart rates of Holstein-Friesian dairy heifers. *Alexandra Green* | Service dogs for epileptic people, a systematic literature review. *Amélie Catala* |
| **11.00** | **Break and posters**, Atlantic Veterinary College ........................................... | |
| 11.45 | For the love of learning: heifers' motivation to participate in a learning task. *Rebecca Meagher* | Stranger-directed aggression in pet dogs: Owner, environment, training and dog-associated risk factors. *Hannah Flint* |
| 12.00 | The effect of feeding and social enrichment during the milk-feeding stage on cognition of dairy calves. *Kelsey Horvath* | The effect of the Danish dangerous dog act on the level of dog aggression in Denmark. *Björn Forkman* |
| 12.15 | Milk feeding strategy and its influence on development of feeding behavior for dairy calves in automated feeding systems. *Melissa Cantor* | The effect of public cat toilet provision on elimination habits of ownerless free-roaming cats in old-town Onomichi, Japan. *Aira Seo* |

| 12.30 | Consistently efficient: the use of feeding behavior to identify efficient beef cattle. *Ira Parsons* | How dependent are ownerless free-roaming cats on water supplied by voluntary cat caretakers in old-town Onomichi, Japan? *Hajime Tanida* |
| 12.45 | Multisensorial stimulation promotes affinity between dairy calves and stockperson. *Mateus Paranhos Da Costa* | Examining relationships between cat owner attitudes and cat owner reported management behaviour. *Grahame Coleman* |

**13.00**  **Box lunch**, Atlantic Veterinary College ......................................................

**14.00–20.00  Excursions and lobster supper/vegan stir fry**, North Rustico

## Thursday August 2nd

|  | Session 7<br>**Veterinary aspects of ethology and welfare**<br>Duffy Science Centre Amphitheatre | Session 8<br>**Broiler and turkey: behaviour and welfare**<br>Alex H. MacKinnon Auditorium, Don and Marion McDougall Hall |
|---|---|---|
| 9.00 | Precision phenotyping as a tool to automatically monitor health and welfare of individual animals housed in groups. *T. Bas Rodenburg* | Assessing positive welfare in broiler chickens: Effects of specific environmental enrichments and environmental complexity. *Judit Vas* |
| 9.30 | The impact of Northern fowl mite infestation (*Ornithonyssus sylviarum*) on nocturnal behavior of White Leghorn laying hens. *Leonie Jacobs* | Does increased environmental complexity improve leg health and welfare of broilers? *Anja Riber* |
| 9.45 | Scratching that itch: Dairy cows with mange work harder to access a mechanical brush. *Ana Carolina Moncada* | Individual distress calls as a flock-level welfare indicator. *Lucy Asher* |
| 10.00 | Automated feeding behaviors associated with subclinical lung lesions in pre-weaned dairy calves. *Caitlin Cramer* | The effect of calcium propionate in feeding strategies for broiler breeders on palatability and conditioned place preference. *Aitor Arrazola* |
| 10.15 | Changes in social behaviour in dairy calves infected with *Mannheimia haemolytica*. *Catherine Hixson* | Hepatic damage and learning ability in broilers. *Alexandra Harlander-Matauschek* |
| 10.30 | Can calving assistance influence newborn dairy calf lying time? *Marianne Villettaz Robichaud* | The importance of habituation when using accelerometers to assess activity levels of turkeys. *Rachel Stevenson* |
| 10.45 | Fitness for transport of cull dairy cows at auction markets in British Columbia. *Jane Stojkov* | Do we have effective possibilities to reduce injurious pecking in tom turkeys with intact beaks? *Jutta Berk* |

# Thursday August 2nd

| 11.00 | **Break and posters,** Atlantic Veterinary College ........................................... |
|---|---|

|  | Session 9 | Session 10 |
|---|---|---|
|  | **Lameness** | **Laying hen behaviour and welfare (a)** |
|  | Duffy Science Centre Amphitheatre | Alex H. MacKinnon Auditorium, Don and Marion McDougall Hall |

| 11.45 | Relationships between type of hoof lesion and behavioural signs of lameness in Holstein cows housed in tie-stall facilities. *Megan Jewell* | Feather pecking phenotype affects behavioural responses of laying hens. *Jerine Van Der Eijk* |
|---|---|---|
| 12.00 | Dairy cows alter their lying behavior when lame during the dry period. *Ruan Daros* | An investigation of associations between management and feather damage in Canadian laying hens housed in furnished cages. *Caitlin Decina* |
| 12.15 | Effects of divergent lines on feed efficiency and physical activity on lameness and osteochondrosis in growing pigs. *Marie-Christine Meunier-Salaün* | Enriching floor pens mitigates effects of extended pullet housing on subsequent distribution of laying hens in an aviary. *Sarah Maclachlan* |
| 12.30 | Osteoarthritis impairs performance in a spatial working memory task in dogs. *Melissa Smith* | Enriching floor pens mitigates effects of extended pullet housing on resource use and activity of individual hens in aviaries. *Janice Siegford* |
| 12.45 | Circadian rhythms in activity in healthy and arthritic dogs. *Jack O'Sullivan* | Too close for comfort? Effects of stocking density on the comfort behaviours of three strains of pullets in enriched cages. *Lorri Jensen* |

| 13.00 | **Lunch,** Wanda Wyatt Dining Hall ....................................................... <br> including **Faculty-Student Lunchtime Session** |
|---|---|

|  | Session 11 | Session 12 |
|---|---|---|
|  | **Pain** | **Laying hen behaviour and welfare (b)** |
|  | Duffy Science Centre Amphitheatre | Alex H. MacKinnon Auditorium, Don and Marion McDougall Hall |

| 14.30 | The assessment and alleviation of pain in pigs - where are we now and what does the future hold? *Sarah Ison* | The effect of rearing environment on laying hen keel bone impacts sustained in enriched colony cages. *Allison Pullin* |
|---|---|---|
| 14.45 | A multimodal approach for piglet pain management after surgical castration. *Abbie Viscardi* | Ramps to facilitate access to different tiers in aviaries reduce keel bone and foot pad disorders in laying hens. *Frank Tuyttens* |
| 15.00 | A systematic review of local anesthetic and/ or systemic analgesia on pain associated with cautery disbudding in calves. *Charlotte Winder* | Startle reflex as a welfare indicator in laying hens. *Misha Ross* |

## Thursday August 2nd

| | | |
|---|---|---|
| 15.15 | Do dairy calves experience ongoing, non-evoked pain after disbudding? *Sarah Adcock* | Human-hen relationship on grower and laying hen farms in Austria and associations with hen welfare. *Susanne Waiblinger* |
| 15.30 | Analgesics on tap: efficacy assessed via self-administration and voluntary wheel-running in mice. *Jamie Ahloy Dallaire* | Laying hen preference for consuming feed containing excreta from other hens. *Caroline Graefin Von Waldburg-Zeil* |

**15.45**      **Break and posters, Atlantic Veterinary College** . . . . . . . . . . . . . . . . . . . . . . . . . . . . . . . . . . . . . . . . . . . . . .

**16.30–17.15**    **ISAE Annual General Meeting**
Alex H. MacKinnon Auditorium, Don and Marion McDougall Hall

**18.30–23.59**   **Conference Dinner, Wanda Wyatt Dining Hall** . . . . . . . . . . . . . . . . . . . . . . . . . . . . . . . . . . . . . . . . . .

## Friday August 3rd

| | Session 13<br>**Play behaviour**<br>Duffy Science Centre Amphitheatre | Session 14<br>**Applied ethology and wildlife management**<br>Alex H. MacKinnon Auditorium, Don and Marion McDougall Hall |
|---|---|---|
| 9.00 | The unity and differentiation in mammalian play. *Marek Špinka* | The Canadian commercial seal hunt. *Pierre-Yves Daoust* |
| 9.30 | Do any forms of play indicate positive welfare? *Georgia Mason* | Animal welfare complementing or conflicting with other sustainability issues. *Donald Broom* |
| 10.00 | Determining an efficient and effective rat tickling dosage. *Megan Lafollette* | Impacts of population management with *Porcine Zona Pellucida* vaccine on the behaviour of a feral horse population. *Abbie Branchflower* |
| 10.15 | Odour conditioning of positive affective states: Can rats learn to associate an odour with tickling? *Birte Nielsen* | Why does annual home range predict stereotypic route-tracing in captive *Carnivora*? *Miranda Bandeli* |
| 10.30 | Chickens play in the wake of humans. *Ruth Newberry* | Effects of manipulating visitor proximity and intensity of visitor behaviours on little penguins (*Eudyptula minor*). *Samantha Chiew* |
| 10.45 | Does the motivation for social play behaviour in dairy calves build up over time? *Maja Bertelsen* | |

**11.00**      **Break and posters, Atlantic Veterinary College** . . . . . . . . . . . . . . . . . . . . . . . . . . . . . . . . . . . . . . . . . . . . . .

# Friday August 3rd

| | Session 15<br>**Swine behaviour and welfare (a)**<br>Duffy Science Centre Amphitheatre | Session 16<br>**Sheep and goat: behaviour and welfare**<br>Alex H. MacKinnon Auditorium, Don and Marion McDougall Hall |
|---|---|---|
| 11.45 | Piggy in the middle: The cost-benefit trade-off of a central network position in regrouping aggression. *Simone Foister* | Improving our understanding of assessing fear in sheep. *Paul Hemsworth* |
| 12.00 | Preferential associations in recently mixed group-housed finisher pigs. *Carly O'Malley* | Effects of flooring surface and a supplemental heat source on dairy goat kids: growth, lying times and preferred location. *Mhairi Sutherland* |
| 12.15 | Estimation of genetic parameters of agonistic behaviors and their relation to skin lesions in group-housed pigs. *Kaitlin Wurtz* | The effect of outdoor space provision on lying behaviour in lactating dairy goats. *Hannah Freeman* |
| 12.30 | Pen-level risk factors associated with tail lesions in Danish weaner pigs: a cross sectional study. *Franziska Hakansson* | Positive handling of goats does not affect human-directed behaviour in the unsolvable task. *Jan Langbein* |
| 12.45 | Tracing the proximate goal of foraging in pigs. *Lorenz Gygax* | Evaluation of alternatives to cautery disbudding of dairy goat kids using behavioural measures of post-operative pain. *Melissa Hempstead* |

**13.00** Lunch, Wanda Wyatt Dining Hall . . . . . . . . . . . . . . . . . . . . . . . . . . . . . . . . . . . . . . . . . . . . . . . . . . . . .
including student-facilitated discussion topics

| | Session 17<br>**Swine behaviour and welfare (b)**<br>Duffy Science Centre Amphitheatre | Session 18<br>**Euthanasia**<br>Alex H. MacKinnon Auditorium, Don and Marion McDougall Hall |
|---|---|---|
| 14.30 | Effects of enrichment and social status on enrichment use, aggression and stress response of sows housed in ESF pens. *Jennifer Brown* | Goat kids display minimal aversion to relatively high concentrations of carbon dioxide gas. *Suzanne Millman* |
| 14.45 | Motivated for movement? A comparison of motivation for exercise and food in stall-housed sows and gilts. *Mariia Tokareva* | A two-step process of nitrous oxide before carbon dioxide for humanely euthanizing piglets: on-farm trials. *Rebecca Smith* |
| 15.00 | Effects of (a switch in) enriched vs. barren housing on the response to reward loss in pigs in a negative contrast test. *Lu Luo* | Aversion to carbon dioxide gas in turkeys using approach-avoidance and conditioned place avoidance paradigms. *Rathnayaka Bandara* |
| 15.15 | Effects of dietary L-glutamine as an alternative for antibiotics on the behavior and welfare of weaned pigs after transport. *Jay Johnson* | Variability in the emotional responses of rats to carbon dioxide. *Lucía Amendola* |

## Friday August 3rd

| 15.30 | Examining the relationship between feeding behaviour and sham chewing in nulliparous group-housed gestating sows. *Rutu Acharya* | Vets, owners, and older cats: Exploring animal behavioural change in euthanasia decision-making. *Katherine Littlewood* |

**15.45**  **Break, Atlantic Veterinary College** ...............................................................

| | **Session 19** **Stereotypic behaviour** Duffy Science Centre Amphitheatre | **Session 20** **Veterinary procedures** Alex H. MacKinnon Auditorium, Don and Marion McDougall Hall |

| 16.30 | Sham chewing and sow welfare and productivity in nulliparous group-housed gestating sows. *Lauren Hemsworth* | A shy temperament correlates with respiratory instability during anaesthesia in laboratory-housed New Zealand white rabbits. *Caroline Krall* |
| 16.45 | Neurophysiological correlates of different forms of stereotypic behaviour in a model carnivore. *María Díez-León* | Are towel wraps a better alternative to full-body restraint when handling cats? *Carly Moody* |
| 17.00 | Boredom-like states in mink are rapidly reduced by enrichment, and are unrelated to stereotypic behaviour. *Andrea Polanco* | Effect of owner presence on dog responses to a routine physical exam in a veterinary setting. *Anastasia Stellato* |

| | **Session 21** **Other** Duffy Science Centre Amphitheatre |

| 17.15 | The ubiquity and utility of camels. *Michael Appleby* |

| 17.20 | **Closing of conference,** Duffy Science Centre Amphitheatre |

| 18.30–23.00 | **Farewell party,** McMillan Hall, W.A. Murphy Student Centre |

# Table of contents

## Poster presentations

## Session 03. Putting ethology into practice

### Oral presentations

### Poster presentations

## Session 04. Laboratory animal: behaviour and welfare

### Oral presentations

### Poster presentations

## Session 05. Cattle behaviour and welfare

### Oral presentations

## Session 06. Companion animal behaviour and welfare

### Oral presentations

**Poster presentations**

## Session 07. Veterinary aspects of ethology and welfare

**Oral presentations**

**Poster presentations**

## Session 08. Broiler and turkey: behaviour and welfare

### Oral presentations

### Poster presentations

## Session 09. Lameness

### Oral presentations

### Poster presentations

## Session 10. Laying hen behaviour and welfare (a)

### Oral presentations

### Poster presentations

## Session 11. Pain

### Oral presentations

### Poster presentations

## Session 12. Laying hen behaviour and welfare (b)

### Oral presentations

**Poster presentations**

## Session 13. Play behaviour

### Oral presentations

## Session 14. Applied ethology and wildlife management

### Oral presentations

## Session 16. Sheep and goat: behaviour and welfare

### Oral presentations

**Poster presentations**

# Session 17. Swine behaviour and welfare (b)

**Oral presentations**

## Session 19. Stereotypic behaviour

## Session 20. Veterinary procedures

## Session 21. Other

**Oral presentations**

### The other-minds problem in other species

*Stevan Harnad*

*Université du Québec à Montréal, Institut des science cognitives, 320, rue Sainte-Catherine Est, Montréal, Québec, H2X 1L7, Canada; harnad@uqam.ca*

Both ethology and comparative psychology have long been focussed on behavior. The ethogram, like the reinforcement schedule, is based on what organisms *do* rather than what they *feel*, because what they do can be observed and what they feel cannot. In philosophy, this is called the "other minds problem": the only feelings one can feel are one's own. The rest is just behavior, and inference. Our own species does have one special behavior that can penetrate the other-minds barrier directly (if we are to be believed): We can *say* what's on our mind, including what we feel. No other species has language. Many can communicate non-verbally, but that's *showing*, not *telling*. In trying to make do with inferences from behavior, the behavioral sciences have been at pains to avoid "anthropomorphism" – the attribution of feelings and knowledge to other species by way of analogy with how we would be feeling if we were doing what they were doing under analogous conditions. Along with Lloyd Morgan's Canon – that we should not make a mentalistic inference when a behavioral one will do – the abstention from anthropomorphism has further narrowed inroads on the other-minds problem. We seek operational measures to serve as proxies for what our own native mammalian mind-reading capacities tell us is love, hunger, fear. Yet our human mind-reading capacities – biologically evolved for care-giving to our own progeny as well as for social interactions with our kin and kind – are really quite acute. That (and not just language) is why we do not rely on ethograms or inferences from operational measures in our relations with one another. But what about other species? And *whose problem* is the "other-minds problem"? Philosophers think of it as our problem, in making inferences about the minds of our own conspecifics. But when it comes to other species, and in particular our interactions with them, surely it is *their* problem if we misinterpret or fail to detect what or whether they are feeling. We no longer think, as Descartes did, that all nonhuman species are insentient zombies, to do with as we please – at least not in the case of mammals and birds. But, as I will show, we are still not sure in the case of fish, other lower vertebrates, and invertebrates. Debates rage; and if we are wrong, and they do feel, the ones suffering a monumental problem are trillions and trillions of fish (and other marine species). Ethologists have been using ingenious means to demonstrate (operationally) what to any human who has seen a fish struggling on a hook (or a lobster in a boiling pot) is already patently obvious. For such extreme cases some have urged inverting Morgan's Canon and adopting the 'Precautionary Principle": giving the other minds the benefit of the doubt unless there is proof to the contrary. Yet even with mammals, whose sentience is no longer in doubt, there are problematic cases. Applied Ethology is concerned with balancing the needs of other species with the needs of our own. Everyone agrees that sentient species should not be hurt needlessly. Most people today consider using animals for food or science a matter of vital necessity for our species. (It is not clear how right they are in the case of food, at least in developed economies, and we all know that not everything done in the name of science is vitally necessary, but neither of these premises will be called into question here.) Entertainment, on the other hand, is clearly not a matter of vital necessity. Yet some forms of entertainment entail hurting other species that are known to be sentient. The rodeo is a prominent example. I will close by describing the results of the analysis of a remarkable database gathered in order to test rodeo practices in Quebec in light of a new law that accords animals the status of "sentient beings with biological needs" who may no longer be "subjected to abuse or mistreatment that may affect their health" (except in the food industry and in scientific research).

## Cognitive evaluation of predictability and controllability and implications for animal welfare

*Caroline Lee and Dana Campbell*

*CSIRO, Agriculture and Food, FD McMaster Laboratory, New England Hwy, Armidale 2350, Australia; caroline.lee@csiro.au*

New systems and technologies for livestock management are increasing in their development and use. One example of these is virtual fencing that is being commercialised and if widely-adopted, will result in increased labour savings, reduced costs of fixed fencing and improved animal monitoring. While animal welfare may improve through increased monitoring, there is also potential for negative impacts if the expectations of the new system do not align with the animals' cognitive and learning abilities. In the example of virtual fencing, the location of cattle and sheep are controlled through avoidance learning so that animals learn to respond to an audio conditioning stimulus in order to avoid receiving an electric shock (response stimulus). If the animal is able to learn the appropriate avoidance behaviour i.e. the cues are predictable (audio warning always precedes a shock) and controllable (responding to the audio cue prevents receiving the shock) then welfare impacts are minimal. Conversely, in the situation where animals do not have the ability to learn the association between the audio and shock cue, they would be in a situation of low predictability and controllability which can lead to states of helplessness or hopelessness with serious implications for animal welfare. For welfare to be acceptable, it is important that all animals in a virtual fencing system are capable of learning the appropriate associations so that they can avoid receiving the shock. The welfare impacts of systems such as virtual fencing depend on the animals' cognitive evaluation of their interaction with the technology and are based on several criteria of which predictability and controllability are central. We will describe a framework for welfare assessment in relation to cognitive evaluation of the predictability and controllability of a situation and discuss implications for animal welfare. The framework is applicable to new technologies or systems that involve animal learning.

**Understanding the bidirectional relationship of emotional and cognitive systems to measure affective state in the animal**

Sebastian McBride
*Aberystwyth University, IBERS, Penglais, Aberystwyth, SY23 3DA, United Kingdom;*
*sdm@aber.ac.uk*

Understanding the emotional consequences of keeping animals within a managed environment is fundamental to determining their welfare status. In recent years, due to an improved understanding of how cognitive and emotional brain systems interact, cognitive measures are now being used as indicators of the emotional state of animals. Most of the recent work has focused on tests of 'judgement bias' whereby animals exposed to either rewarding or stressful stimuli tend to be more or less optimistic (respectively) about the outcome of ambiguous cues. This judgement bias approach has been used convincingly to assess the effects of husbandry situations and events on generalised affective state in a range of mammal and bird species. Whilst such approaches have been extremely useful, neurophysiological evidence suggests that, given the complexity of the bidirectional relationship between emotional and cognitive system, there exists an opportunity to use alternative cognitive paradigms to probe additional dimensions of affective state. These additional dimensions may be extremely pertinent to the assessment of the welfare of the animal. Alternative cognitive tasks to probe animal affective state are presented here using examples from two recently developed operant cognitive devices for sheep and horses. The tasks in question are the two-choice visual discrimination reversal task, a visual attention task, the stop-signal reaction time task, the 3 choice serial reaction time task and the delayed match to sample task. We present data from these cognitive tasks (using sheep and horses) and discuss: (1) the brain mechanisms and systems by which they are mediated (e.g. dopaminergic modulation of cortico-striatal loops by mid-brain systems, bottom-up bias of fronto-parietal networks mediated though limbic and limbic system modulation of hippocampal spatial function); (2) how these brain mechanisms are modulated by current and historic stress events as well as the emotional state of the animal, to produce differences in cognitive performance. Through this analysis, potential cognitive tasks (additional to the judgement bias task) for the measurement of animal welfare are identified and discussed.

## Evaluation of sheep anticipatory behaviour and lateralisation by means of fNIRS technique

*Matteo Chincarini[1], Lina Qiu[2], Michela Minero[3], Elisabetta Canali[3], Clara Palestrini[3], Nicola Ferri[4], Lorenzo Spinelli[2], Alessandro Torricelli[2,5], Melania Giammarco[1] and Giorgio Vignola[1]*
*[1]Università degli Studi di Teramo, Fac. di Medicina Veterinaria, S.S. 18, Loc. Piano d'Accio, 64100 Teramo, Italy, [2]Politecnico di Milano, Dip. di Fisica, Piazza Leonardo da Vinci, 32, 20133 Milano, Italy, [3]Università degli Studi di Milano, Dip. di Medicina Veterinaria, Via Giovanni Celoria, 10, 20133 Milano, Italy, [4]Istituto Zooprofilattico Sperimentale dell'Abruzzo e del Molise G. Caporale, via Campo Boario, 64100 Teramo, Italy, [5]Consiglio Nazionale delle Ricerche, Istituto di Fotonica e Nanotecnologie, Piazza Leonardo da Vinci, 32, 20133 Milano, Italy; mchincarini@unite.it*

Lateralisation has been widely studied in animals to investigate their emotional state. Here we applied functional near-infrared spectroscopy (fNIRS) to assess brain lateralisation on 15 free moving sheep during anticipation of two events with different emotional valence. The fNIRS technique relies on the neurovascular coupling mechanism and on the ability of near-infrared light to penetrate deeply in biological tissue, exploiting the different absorption spectra of $[O_2Hb]$ and $[HHb]$ in brain cortex. Animals can be easily habituated to wear fNIRS probes for short periods of time (e.g. 20 minutes). The study was approved by the Ministry of health (authorisation no. 457/2016-PR) according to EU Directive 2010/63 on the protection of animals used for scientific purposes. We trained the sheep with a classical conditioning paradigm to anticipate either a positive or an adverse event after the presentation of acoustic and visual stimuli. The positive event consisted of a bucket filled with food, while the adverse event consisted of an empty bucket. In order to reach the bucket, sheep had to detour an obstacle indicating a side preference. Only 8 sheep successfully completed the training. Behavioural results were analysed with Wilcoxon signed rank test, and a laterality index (LI) for the side used to detour the obstacle was calculated. Changes in $[O_2Hb]$ and $[HHb]$ were analysed using student's t-test. Sheep detoured the obstacle more often at the right side in both training and test phase (94% and 85%; $P<0.01$), regardless the valence of the trials. After the stimuli, sheep were more active ($P \leq 0.001$) and focused on the gate ($P \leq 0.001$). A higher level of activity was observed during the anticipation of the adverse event ($P<0.05$) compared to the positive. During the anticipation phase, we observed a neural activation in the cortex as measured by an increase of $[O_2Hb]$ and a decrease of $[HHb]$ ($P<0.001$). A significant difference in $[O_2Hb]$ was found between the two hemispheres ($P<0.001$), with the right side showing a larger increase in anticipation of both events. The combination of behavioural and brain indicators allows us to conclude that sheep learned the association between stimuli and an upcoming event. However, we found no clear indication that sheep perceived the valence of the two events differently. Even though further evaluation on a larger animal sample and different contexts is required, the present results suggest that fNIRS may offer a valuable tool to get an insight of brain functionality associated with upcoming events. This work was supported by MIUR-PRIN2015 (Grant 2015Y5W9YP).

## Assessing differential attention to simultaneous negative and positive video stimuli in sheep: moving toward attention bias

Camille M.C. Raoult[1,2] and Lorenz Gygax[3]
[1]Veterinary Public Health Institute, University of Bern, Vetsuisse Faculty, Animal Welfare Division, Länggassstrasse 120, 3012 Bern, Switzerland, [2]FSVO, Centre for Proper Housing of Ruminants and Pigs, Tänikon 1, 8563 Ettenhausen, Switzerland, [3]ADTI, Humboldt-Universität zu Berlin, Unter den Linden 6, 10099 Berlin, Germany; camille.raoult@agroscope.admin.ch

In humans, negative mood leads to heightened attention towards negative compared with positive stimuli. Therefore, testing attention bias could potentially be used to assess mood in animals. This approach may be easier in comparison with a cognitive bias test because it takes advantage of an animals' spontaneous reaction and does not depend on a lengthy training phase. In order to develop an attention bias test, we conducted a study to assess differential attention to simultaneous negative and positive video stimuli in sheep. Two silent videos showing either predators as a presumed negative or conspecifics as a positive stimulus were projected simultaneously in about natural size. 28 sheep were tested individually, once a day on four days, while being restrained in a feed station. Each session was composed of 18 trials of 30 s each with inter-trial durations of 15-21 s. The content of the videos was manipulated by varying stimulus intensity (i.e. number of animals and type of behaviour) and distance (i.e. animals presented in foreground, middle ground or background). Moreover, each sheep experienced the negative stimuli on the right and left side. Sheep were habituated either 0, 5, or 9 times to the experimental set up. The sheep's head and ear positions and movements were automatically tracked and the frontal cortical brain activity was recorded using functional near-infrared spectroscopy. Linear-mixed effects models were used and a final model was chosen based on BIC (model probability, mPr; evidence ratio with the null model, $E_0$). Neither sheep's head and ear positions and movements nor frontal cortical brain activity varied according to the type of stimuli shown (mPr>0.84; $E_0$>6). However, we found a clear phase effect (30 s stimulus vs 6 s pre- and post-stimulus each): the deoxyhemoglobin concentration increased in the first third of the stimulus phase. The stimulus phase was also characterized by more head movements, more switches of the head from one side to the other, more ear movements, a lower proportion of time with asymmetrical ears and a higher proportion of time with both ears forward. Habituated sheep had their ears forward in a larger proportion of stimulus phase (model estimate and [95% CI]: 0.997 [0.992; 0.998]) than unhabituated sheep (0.61 [0.38; 0.81]). The inverse pattern was observed with the proportion of stimulus phase with passive ears (habituated: 0.009 [0.004; 0.02]; unhabituated: 0.29 [0.15; 0.47]). We had expected that sheep's head and ear positions and movements, as well as their frontal cortical brain activity, would vary according to the valence, intensity and distance of the two simultaneously presented stimuli. However, we were not able to support this notion with the present experimental set-up. Nevertheless, sheep clearly reacted to the presentation of the stimuli indicating that they paid attention to these. In addition, habituated sheep may have been focused even more on the videos than unhabituated sheep.

### An acoustic indicator of positive emotions in horses?

*Mathilde Stomp, Maël Leroux, Marjorie Cellier, Séverine Henry, Alban Lemasson and Martine Hausberger*
*Université de Rennes, Ethology_UMR6552, Station biologique de Paimpont, Beauvais, 35380 Paimpont, France; mathilde.stomp@univ-rennes1.fr*

Assessing positive emotions is crucial for identifying how animals perceive the conditions offered. However, indicators of positive emotions are still scarce and many proposed behavioural markers have proven ambiguous. Studies have established a link between acoustic signals and emitter's internal state, but few related to positive emotions and still fewer considered non-vocal sounds (e.g. purring in felids). One of them, the snort, is shared by several perrisodactyls and has been associated mainly with positive contexts. We hypothesized that this could be also the case in horses. These vibrating sounds produced by nostrils during expiration are often confused with the blows, another sound of exhalation, but more intense and of a shorter duration. Blows have been shown to be emitted in an alarm context while snorts have up to now been considered as having a hygienic function. However, pilot observations have revealed that snorts were produced more in some individuals than in others, despite identical air conditions. Here, we hypothesized that snorts in healthy animals might reflect the animal's psychological state, following a physiological change related to mild positive excitations. We recorded 48 horses belonging to four different populations: two in facilities offering restricted life conditions (e.g. single stall), two living in facilities with more naturalistic conditions (e.g. stable groups in pastures). The immediate context of production (e.g. stall/pasture) and the horse's activity (e.g. eat) and state (ears position) were observed using the focal behavioural sampling method. We additionally performed an evaluation of the welfare state, using common indicators (i.e. stereotypies, aggressiveness toward human, ears backwards while feeding, prolonged orientation toward a wall in stall). A total chronic stress score (TCSS) was calculated per horse from the data obtained for each of these indicators, reflecting how much the welfare state was altered. The results showed that: (1) snort production was significantly associated with positive situations (e.g. while feeding: in stall (hay): 67.3% of the snorts; in pasture (grass): 69.6% of the snorts), and horses expressed more ears in forward or sidewards positions (reflecting more positive internal states) while snorting compared to a basal situation (Wilcoxon test, $Z=3.47$, $P<0.001$); (2) the horses living in restricted conditions produced twice as many snorts when in pasture compared to the stall condition (Wilcoxon test, $Z=2.84$, $P=0.004$); (3) the naturalistic population emitted significantly more snorts than restricted conditions ones in comparable contexts (Kruskal-Wallis test, $\chi^2=12.851$, df=2, $P=0.001$); (4) the frequency of snorts was especially negatively correlated with the TCSS (reflecting compromised welfare): the lower the TCSS, the higher the snort rate (Spearman correlation, $r=-0.39$, $P=0.005$). These four results converge to indicate that snorts could reflect a positive emotional state. Snorts could be a useful tool for improving management practices, based on the horse's subjective assessment of the situations.

## Sleep vs no sleep; equine nocturnal behavior and its implications for equine welfare

*Linda Greening*
*University Centre Hartpury, Hartpury College, Hartpury House, GL193BE, Gloucester, United Kingdom; linda.greening@hartpury.ac.uk*

Sleep is well recognized as an important facilitator of physical well-being and optimal mental functioning. Providing nocturnal environments that promote sleep should therefore be considered a key responsibility of human care givers for animals managed within artificial settings. Stabling is a convenient and widely used method of managing the domestic horse, with a number of human-perceived benefits for the horse. Evidence exists, highlighting how intrinsic diurnal behavioural patterns are limited due to this type of management, including grazing, social interaction, and free-movement. This abstract outlines a series quasi-experimental and observational studies completed over the last five years, which describe and provide insight into equine nocturnal behavioural patterns. All studies employed focal continuous sampling methods and involved the use of infrared CCTV systems. These studies conform to ISAE Ethical Guidelines, with ethical approval gained from the Hartpury Ethics Committee. Results highlight that nocturnal behaviour is indeed affected by a range of factors linked to stabling. In the first study published in 2013, duration of nocturnal recumbent and ingestive behaviours were recorded from 7 pm to 7 am for stabled horses bedded on either wood shavings (n=5) or straw (n=5). Differences between groups were analysed using Mann Whitney-U and although non-significant ($P>0.05$), horses bedded on straw spent a greater proportion of the nocturnal behavioural time budget recumbent (29%) compared with horses bedded on shavings (12%). Stable designs enabling social interaction through the use of barred instead of brick walls are generally considered an example of good welfare. After observing twelve horses for two consecutive nights between 10 pm and 7 am, significantly less (t=2.436, $P<0.05$) standing sleep was recorded for horses stabled in solid walled, and although non-significant a greater proportion of time was spent recumbent compared to horses in barred wall stables. Ten horses observed for a total of four non-consecutive nights between 7 pm and 7 am, following the move from overnight turnout to overnight stabling demonstrated an acclimatization period through a significant increase ($P<0.01$) in duration of recumbent behaviour from the first week of stabling (5% of the time budget) to six weeks later (16% of the time budget). A preliminary study investigating whether sleep-related behaviours impact upon subsequent ridden performance found a positive correlation ($R^2=0.513$) between competition performance and average duration of nocturnal recumbent behaviour. Anecdotally horse owners may not consider what the horse does whilst stabled overnight, with assumptions about the occurrence of sleep. This collection of studies highlights how husbandry practices may be (re)considered to facilitate sleep-related behaviours. Moreover, although tenuous, links to performance are also evident, therefore next steps might involve determining specifically how lack of sleep manifests within diurnal equine behaviour.

**Shelter-seeking behaviour in domestic donkeys and horses in a temperate climate**

*Britta Osthaus[1], Leanne Proops[2], Sarah Long[3], Nikki Bell[3], Kristin Hayday[3] and Faith Burden[3]*
*[1]Canterbury Christ Church University, School of Psychology, Politics and Sociology, North Holmes Road, CT1 1QU Canterbury, United Kingdom, [2]University of Portsmouth, Centre for Comparative and Evolutionary Psychology, Department of Psychology, PO1 2DY Portsmouth, United Kingdom, [3]The Donkey Sanctuary, Research Centre, EX10 0NU Sidmouth, United Kingdom; britta.osthaus@canterbury.ac.uk*

Donkeys and horses differ substantially in their evolutionary history, physiology, behaviour and husbandry needs. Donkeys are often kept in climates that are colder and wetter than those they are adapted to and therefore may suffer impaired welfare unless sufficient protection from the elements is provided. We compared the shelter-seeking behaviours of donkeys and horses in relation to temperature, precipitation, wind speed and insect density. Our study collected 13,612 day-time data points (location of each animal, their activity such as feeding, resting, moving, etc., and insect-related behaviours) from 75 donkeys and 65 horses (unclipped and un-rugged) with free access to man-made and natural shelters between September 2015 and December 2016 in the South-West of the UK. Each animal was observed at least once a week, with an average of 65 observations per individual overall. Even though the UK climate is quite mild (1 to 33 degrees Celsius in our sample), the preliminary results showed clear differences in the shelter seeking behaviour between donkeys and horses. Overall donkeys were observed far more often inside their shelters than horses ($\chi^2(1)$=1,783.1, $P<0.001$). They particularly sought shelter when it was raining: there was a 54.4%-point increase (35 to 89.4%) in the proportions of donkeys sheltering in rainy conditions, in comparison to a 14.5%-point increase in horses (9.6 to 24.1%). Results of binary logistic regressions indicated that there was a significant association between species, precipitation and shelter-seeking behaviour ($\chi^2(3)$=2,750.5, $P<0.001$). Horses sought shelter more frequently when it got hotter, whereas donkeys sought shelter more often in colder weather ($\chi^2(3)$=2,667.3, $P<0.001$). The wind speed (range 0 to 8 m/s – calm to moderate breeze) had an effect on location choice, and this again differed significantly between donkeys and horses ($\chi^2(3)$=1,946.5, $P<0.001$). In a moderate breeze, donkeys tended to seek shelter whereas horses moved outside. The insect-related behaviours were closely related to temperature and wind speeds. The donkeys' shelter-seeking behaviour strongly suggests that in temperate climates they should always have access to shelters that provide sufficient protection from the environment.

### Social referencing in the domestic horse

*Christian Nawroth[1], Marie-Sophie Single[2] and Anne Schrimpf[3]*
[1]*Leibniz Institute for Farm Animal Biology, Institute of Behavioural Physiology, Wilhelm-Stahl-Allee 2, 18196 Dummerstorf, Germany,* [2]*Technical University of Munich, Physiology Weihenstephan, Weihenstephaner Berg 3, 85354 Freising, Germany,* [3]*Max Planck Institute for Human Cognitive and Brain Sciences, Department of Neurology, Stephanstraße 1a, 04103 Leipzig, Germany; nawroth.christian@gmail.com*

A detailed comprehension of the socio-cognitive abilities of domestic animals is necessary to improve their management and husbandry in the long term. Domestic companion animals, such as dogs, are able to use human emotional information directed to an unfamiliar object or other individuals to guide their behaviour, a mechanism known as social referencing. However, it is not clear whether other domestic species show similar socio-cognitive abilities in interacting and communicating with humans. We investigated whether horses (n=46, 1-26 years, 18 females, 25 geldings, 3 stallions) use emotional information provided by humans to adjust their behaviour to a novel object (a plastic bin covered with a shower curtain; height: 120 cm, diameter: 70 cm). Horses received one single test trial (duration 40-120 s) and were randomly assigned to one of two groups: an experimenter was positioned in the middle of a round test arena (diameter: 18 m) and directed her gaze and voice towards the novel object with either: (1) a positive emotional message (excited facial expression, positive vocalization: 'Das ist ja toll' ('This is great'), repeated every 10 s; n=22); or (2) a negative emotional message (anxious facial expression, negative vocalization: 'Das ist ja schrecklich' ('This is terrifying'), repeated every 10 s; n=24). At the beginning of each test trial, the experimenter led the horse towards a starting point in the middle of the arena, with the object being positioned approximately 7 m away from both. Duration of subjects' relative position to the experimenter and the object in the arena (behind experimenter and close to the exit gate, abreast to experimenter, in front of experimenter), as well as frequency and duration of gazing behaviour and physical interactions (towards/with either object or experimenter) were scored and analysed. We found that horses that received a negative emotional message spent less time in front of the experimenter (i.e. directed towards the object; mean % of trial duration ± SE: 17.8±5.4 and 39.5±8.2, respectively; Mann-Whitney U-test; U=353, P=0.043) compared to the positive group. Subjects from the negative group also gazed more often at the object compared to subjects from the positive group (mean gazes per minute ± SE: 3.52±0.26 and 2.53±0.24, respectively; Mann-Whitney U-test; U=145, P=0.009), indicating increased vigilance behaviour. No differences in horses' physical interaction with the experimenter or the object as well as their general activity pattern (locomotion and resting) were found. Our results provide partial evidence that horses use emotional cues from humans to guide their behaviour towards novel objects, while alternative explanations such as a learned response or a general increase in vigilance behaviour cannot be ruled out. These findings have the potential to improve horse handling practices by using these mechanisms to facilitate habituation processes to new environments.

## Irish Traveller horse owners' recognise the importance of natural behaviour to the welfare of horses

*Marie Rowland[1,2], Melanie Connor[1] and Tamsin Coombs[1]*

[1]*SRUC, Roslin Institute, Easter Bush, Midlothian, EH25 9RG, United Kingdom,* [2]*University of Edinburgh, Royal (Dick) School of Veterinary Studies, Easter Bush, Midlothian, EH25 9RG, United Kingdom; marie.rowland@sruc.ac.uk*

Horses are an important aspect of Travellers lives with horse ownership considered one of the last links to their nomadic way of life. Travellers are a distinct ethnic minority group, indigenous to Ireland. Traveller horse owners are often singled out as the main contributors to reduced horse welfare with fly-grazing (unauthorised grazing on private land), tethered and abandoned horses regarded as Traveller practices. Their viewpoint and approach to horse care, management and welfare however is rarely studied. This study explored 14 Irish Traveller horse owners' attitudes to horse care, management and welfare. Two qualitative research methods were used; face-to-face semi-structured interviews and discussion groups. Participants were invited to reply to open-ended questions on the subjects of horse health, feeding, exercise, sentience and social interactions. To obtain key information and define themes present in the transcripts, coding and thematic analysis was performed using Nvivo 10. Coding was done by examining interview transcripts and assigning codes for attitudes, values, meanings, beliefs, experiences and knowledge characteristic of the subject studied. Thematic analysis revealed key concepts that consistently arose and were relevant to the project's aims. The final selection of themes and sub-themes were grounded in these aims and in the views and experience of the participants. Themes were found to be particularly rich in relation to participants understanding of the natural behaviour and natural environment of the horse and its application in their management practices, an approach currently gaining recognition in equine studies. Participants reached a high degree of consensus on the importance of natural behavioural expression with horses considered to be highly social animals. There was unanimity across all participants on the principal of keeping horses outdoors and in herds and 10 participants considered exercise to be important, claiming that it is natural to outdoor horses. Constant access to grazing was deemed essential by 10 participants, 8 stated that rugs/blankets were only necessary in exceptional circumstances while 9 participants regarded the provision of adequate shelter essential. Additionally, the overwhelming perception amongst all participants was that horses have feelings with 11 participants maintaining that horses develop bonds with their conspecifics and 8 claimed that horses bonded with humans. Seasonal feeding patterns, opportunity for social bonding, play and mutual grooming behaviour were further natural practices advocated by this cultural group. In conclusion, it is evident that Traveller horse owners' attitudes and approach to horse care and management in this study supports the natural behaviour of the horse, therefore has the potential for a high level of horse welfare. The findings make a valuable contribution to the under-researched area of Traveller horse welfare. Developing these findings further with larger sample groups is recommended.

**What a horse needs: examining Canadian equine industry participants' perceptions of welfare**

*Lindsay Nakonechny[1,2], Cordelie Dubois[1,2], Emilie Derisoud[3] and Katrina Merkies[1,2]*
[1]*Campbell Centre for the Study of Animal Welfare, 50 Stone Rd E, N1G 2W1 Guelph, Canada,* [2]*University of Guelph, Animal Biosciences, 50 Stone Rd E, N1G 2W1 Guelph, Canada,* [3]*Agrocampus Ouest, 65 Rue de Saint-Brieuc, 35000 Rennes, France; kmerkies@uoguelph.ca*

The Canadian equine industry is made up of a diverse population of individuals involved with horses in a variety of ways further divided by riding discipline, making it difficult to determine baseline attitudes and beliefs toward management and welfare. Understanding perceptions is important when dealing with topics such as welfare as it helps direct educational outreach programs and focus future research. The purpose of this study was to determine stakeholder attitudes on welfare and management in the Canadian equine industry through an online survey distributed via equine organization mailing lists and social media to a broad range of Canadian equine enthusiasts (n=901). Participants were asked to rank what they believed was most important to horses (e.g. companionship) and to identify aspects of the best management environment (e.g. ideal turnout) from a selection of provided answers. They were also asked about undesirable behaviours (e.g. stereotypies, for which descriptions were provided) and their perception of the state of equine welfare within Canada through open-ended comments. Data were examined using descriptive statistics. Participants believed that the opportunity to consume forage (48%) was more important to horses than the opportunity to sleep or search for food (41 and 40% respectively). Participants indicated that the best stall environment should include the ability to see other horses (90%), and access to full-day turnout (87%) in a paddock containing grass, a mud-free area, shelter, hay, companionship, water, and access to salt (>70% for all paddock variables). Although participants felt horses should be fed a diet of primarily forage (63%), there was no clear agreement if concentrates and supplements could be used to meet increased nutritional demands (e.g. due to workload). The majority of participants (89%) felt that stereotypies were not indicative of poor welfare and that undesirable behaviours performed when ridden or driven were primarily the result of poor cueing by riders (67%). Greater than 89% of participants strongly agreed that horses were capable of feeling pain, fear and boredom. Participants indicated the best ways to assess welfare were vital signs (26%), absence of abnormal behaviour (23%) and evaluation of time budgets (21%). There was no consensus as to which horses were at risk for welfare concerns (i.e. specific sectors or disciplines), what the best way to address welfare concerns was and who should be responsible for delivering equine-based education programs to address welfare. The recurring theme of owners' lack of knowledge appeared in discussions of welfare issues within the industry. Overall, participants in the Canadian equine industry agree that there are welfare issues within their industry, but amalgamating their opinions of how to assess and remedy these issues continues to be a challenge as individual perceptions are diverse and divided.

**Equine heart rate variability measures: are they stable from day to day and during different activities?**

*Laurie McDuffee[1], Mary McNiven[1], Molly Mills[1] and William Montelpare[2]*
*[1]Atlantic Veterinary College, Health Management, 550 University Ave, Charlottetown, C1A 4P3, Canada, [2]University Of Prince Edward Island, Applied Human Sciences, 550 University Ave, Charlottetown, C1A 4P3, Canada; lmcduffee@upei.ca*

Heart rate variability (HRV) measures obtained from inter-beat intervals are considered a valuable non-invasive tool for assessment of animal stress, pain and welfare including stress in horses in various housing or husbandry conditions. We plan to use HRV measures to evaluate stress and pain in horses admitted to a veterinary teaching hospital. The reliability of HRV measurements is critical when using the data to interpret stress, pain and welfare of horses. However, the statistical stability of HRV from horses in various circumstances is not well documented. The purpose of this study was to determine the day-to-day stability of HRV measures in stabled horses during three routine activities. Time and frequency domain measures of HRV were obtained on two days for horses of various breeds, sex, and age while confined to a box stall (n=32) and cross ties (n=32), or during walking (n=15). All horses were adults used for riding or racing. HRV data was collected with the Polar® Equine V800 Heart Rate Monitor and analyzed with Kubios® HRV software. R-R interval data was recorded for 10 minutes during the activities, which took place in the horses' normal environment. Five-minute segments were used for analysis, and the medium correction factor was applied. Statistical stability was determined using a test-retest approach to estimate the intraclass correlation coefficients for HRV within the time domain and auto regressive method of frequency domain analyses. The SAS™ PROC Mixed procedure was used to compute ICCs for all horses during all activities, and a subset of horses with 15% or fewer corrected beats. A further subset of horses with ≤10% corrected beats was analyzed for walking. The ICC for horses in box stalls showed moderate to almost perfect agreement (0.5-0.9) for all measures except SDNN (poor agreement: 0.2). For horses confined to cross ties, ICCs showed moderate to almost perfect agreement (0.5-0.9) for all measures except SDNN and NN50, which had fair agreement. The ICC based on walk data showed poor agreement (<0.2) when analysis included >10% corrected beats. For ICCs based on ≤10% corrected beats, SDNN, NN50 improved to moderate agreement while LFpwr n.u., HFpwr n.u., LF/HF showed strong agreement. Previous research showed that HRV reliability decreases with increased movement. Our study indicated that although HRV measures during stall and cross tie confinement were statistically reliable showing day-to-day agreement, HRV measures for individual horses during walking were not reliable under all circumstances. Removing horses that required more than 15% beat correction improved ICCs for stall and cross tie confinement. However, only using horses that met the 10% beat correction improved reliability for the walk activity. The results support the use of HRV as a stable indicator of neurophysiological response during stall and cross tie confinement, but more scrutiny is needed when applied to walking activity among equids.

**Crossing the divide between academic research and practical application of ethology on commercial livestock and poultry farms**

*Temple Grandin*
*Colorado State University, Dept of Animal Science, 650 West Pitkin Street, Fort Collins Colorado 80523-1171, USA; cheryl.miller@colostate.edu*

There are many young managers in commercial animal agriculture in the United States, United Kingdom, and other countries, who do not know what animal ethology is. They often have an agricultural degree with no training in animal behavior. Some have no idea that scientists have already conducted many research studies on animal behavior. How do we cross this divide? It will require three steps: The first step is teaching basic animal behavior to veterinary and animal science undergraduate students. Recently I communicated directly with students from a major UK veterinary college, who had never been taught basics about bull safety. The most basic information that should be taught to undergraduates is: (1) Behavioral principles of livestock handling; (2) importance of good stockmanship to improve animal productivity; (3)principles of animal learning; (4) bull, ram and boar safety; (5) the importance of behavioral needs and environment enrichment; (6) how to recognize abnormal behaviors; and (7)formation of dominance hierarchies. This material should be in introductory courses with practical explanations about why it is important. For example, a nutritionist needs to understand how dominance behavior may reduce access to feed. When I communicate directly with students, they are eager to learn about behavior. The second step is that researchers must communicate with producers in jargon-free language. The third step is training graduate students for management jobs on farms or research careers in industry. In the developed countries, there is a shortage of academic positions for new Ph.D.'s. Graduates in animal behavior subjects can have excellent careers outside of academia. Their training in animal behavior may influence the policies of their employers to improve animal welfare. There are also factors that in the future may block free flow of scientific information. Unfortunately, some research results remain proprietary commercial industry information and they are not published in the scientific literature. To promote the spread of knowledge, academic researchers should avoid signing long-term non-disclosure agreements with industry. These agreements may block scientific publications. Everybody in the field of animal behavior needs to communicate outside their field and explain why behavior is important.

## Integrating ethology into actual practice on Canadian farms

*Jacqueline Wepruk*
*National Farm Animal Care Council, Box 5061, T4L 1W7, Canada; nfacc@xplornet.com*

The National Farm Animal Care Council (NFACC) oversees the process by which Canadian Codes of Practice for the care and handling of farm animals are developed. Canada currently has fourteen Codes guiding on-farm animal care for various types of livestock and poultry, and one Code for livestock and poultry transportation. Animal welfare science, including the field of ethology, has played an important role in how Canada's Codes are developed. Notably, a Scientific Committee reviews the research available on key priority animal welfare issues for the species being addressed. The resulting peer-reviewed report provides valuable information to the Code Committee in developing or revising a Code of Practice. The selection process for the Scientific Committee stipulates the inclusion of a diversity of scientific expertise, including ethologists. In addition, members of the Scientific Committee are tasked with considering research available for each priority welfare issue based on three general overlapping types of concern: biological functioning (including health and productivity), how the animal 'feels' (including measures of pain and preferences), and the naturalness (including the animals ability to perform behaviours that are important to the animal). This directive is based upon David Fraser and colleagues' well known paper on three types of animal welfare problems that cause three types of ethical concerns. The authors of the paper suggest that these problems and concerns should define the subject matter of animal welfare science. This approach has impacted Canada's Codes in a number of ways, notably the inclusion of a blend of requirements and recommended practices based on biological functioning, affective states, and an animal's need to perform certain behaviours. These national Codes now form the basis of industry developed animal care assessment programs, further entrenching the consideration of ethology into on-farm practices. Examples of how animal behaviour has been utilized within Codes and broader on-farm application will be covered. In addition, the presentation will touch upon research that has been conducted on the uptake of the Codes by stakeholder groups as well as the legal standing of Codes in Canada.

## Balancing outcome-based measures and resource-based measures in broiler welfare policies: the NGO perspective

*Monica List*
*Compassion in World Farming, 125 E Trinity Pl, Ste 206, 30030, Georgia; monica.list@ciwf.org*

Welfare assessment in farm animals has shifted to animal-based measures (ABM), specifically outcome-based measures (OBM), which may more accurately reflect the welfare state of the animal, and align with widely accepted definitions of animal welfare as an animal-centric concept. This trend toward OBM welfare assessment is also evident in recent updates to some corporate animal welfare policies. On the other hand, animal protection organizations continue to lobby for changes that ensure what they believe to be critical improvements to environments and practices in farm animal production. The focus of these efforts continues to be on resource-based measures (RBM). In the United States, a coalition of animal protection organizations produced a joint statement on broiler welfare issues in 2017. The statement covers five key issues, identified by the coalition as priorities for the improvement of broiler welfare: (1) a switch to genetics with measurable, improved welfare outcomes; (2) a reduction in stocking density to 6 lb per square foot (30 kg per square meter); (3) environmental modifications including adequate litter, lighting, and provision of enrichment items; (4) adoption of multi-step controlled atmosphere stunning systems for processing; and (5) the use of third party auditing to demonstrate compliance. These key issues were selected based on their relevance to broiler welfare as determined by scientific research, and because they were the minimum common denominator for the groups involved. Furthermore, the joint approach was considered best from a strategic perspective, as it presented a unified front, and minimized the number of different requests being made to food businesses. This work presents the rationale behind the selection of the key issues, and elaborates on how they can be used together with OBM welfare assessment. It also presents a preliminary set of criteria for selecting OBM based not only on their potential to improve welfare, but also on their aptness for comprehensive welfare assessment that also considers RBM and management based measures.

## A global perspective on environmental enrichment for pigs: how far have we come?

*Heleen Van De Weerd*
*Cerebrus Associates Ltd., The White House, 2 Meadrow, Godalming, Surrey GU7 3HN, United Kingdom; heleen@cerebrus.org*

Improvements in the welfare of farmed animals can be achieved through policy formulation, litigation, consumer and NGO pressure, agri-sector business innovation and business engagement, for example via Corporate Social Responsibility. Scientific evidence is an essential foundation for all routes. This presentation will review progress towards the provision of pig enrichment that satisfies pig-specific needs, as a way to enhance welfare. It will cite examples from the global pig sector, draw on pig science and illustrate the activities of animal welfare NGOs. There is a mismatch between the scientific understanding of what constitutes good and effective enrichment for pigs and what happens in practice on pig farms around the world. A review of the situation in the three largest global pork producing regions, highlights that providing enrichment to pigs is a legal requirement in the EU (Council Directive 2008/120/EC). In practice, producers do not always provide enrichment and if they do, it often is a simple object which has no enduring value to the animal. In China, several progressive businesses are providing enrichment to their pigs, but there are currently limitations to what they can achieve. Concerns about bio- and food security severely restrict the use of substrates, there is a low awareness of animal welfare, and staff often have low technical skills. In the USA, farm animal welfare advances have been achieved through legal routes. However, pig enrichment has not yet featured on the political agenda, nor appeared on farms, other than on those operating under certain farm assurance schemes. The main barriers to progress in practice are the inability or unwillingness to invest in animal welfare if there is no obvious return on investment. For enrichment, this means that there is hardly any investment in good innovative objects or in staff time to understand, manage and maintain enrichment. Progress is further thwarted by a lack of knowledge on pig-specific behaviours and on how motivated behaviour can be channelled into adverse behaviour in restricted barren environments. Furthermore, knowledge is limited on how this can be ameliorated with welfare strategies. Suitably skilled people can assist with effective knowledge transfer to practice, despite knowledge gaps on specific enrichment topics (for example on enrichment for sows in groups). Novel drivers of change, such as public business benchmarking can help to incentivise actions, albeit not without business engagement to guide towards realistic goals. High standards of animal care are linked to higher pig welfare, high health status and production benefits. Herein lies an important role for applied ethologist to describe the economic benefits of higher welfare strategies (or adversely, the costs of low welfare). Such knowledge will help to break down the barrier of a lack of investment in enrichment innovation.

## Establishing science-based standards for the care and welfare of dogs in US commercial breeding facilities

*Candace Croney and Traci Shreyer*
*Purdue University, Comparative Pathobiology, 625 Harrison St, 47907, USA; ccroney@purdue.edu*

High demand for purebred dogs in the USA is met in part by commercial dog breeding. Commercial breeding (CB) of dogs, often referred to as "puppy mill" breeding, currently faces public scrutiny and criticism resembling that directed at intensive production of food animals. Although USDA-licensed commercial breeders are regulated and inspected for compliance with federal and state standards of care for dogs, significant concerns remain about the welfare of dogs and puppies from CB facilities. These include maintenance of the dogs in sub-optimal housing without appropriate attention to their behavioral, psychological and physical health needs. While legislating against CB operations appears to offer a solution, it does not address some welfare risks. For example, demand for purebred dogs that cannot adequately be met by small-scale breeders, shelters or rescue groups in the USA increases the likelihood of importing animals, including from countries with unknown welfare standards. Development of care standards for breeding dogs offers a means by which to facilitate dog welfare, public assurance of best practice, and sustainable pet ownership. Comprehensive standards for CB dog care and welfare were drafted and pilot tested at Purdue University, Indiana, USA . Unlike existing US laws and breed club standards, emphasis was placed on dog behavioral well-being given reports that dogs from CB kennels display severe, persistent behavioral and psychological problems, as well as health problems. Dog health, nutrition, genetics, breeding ages, litter limits, euthanasia criteria, retirement and rehoming were also addressed. The standards were pilot-tested at 18 CB kennels located in various Midwestern US states and ranging in size from 15 to 125 breeding bitches. An educational program to support participating dog breeders was created. A tool was developed and used to benchmark the immediately observable welfare status of dogs at test-pilot sites. Physical metrics included body condition score, coat condition, cleanliness, and visual evidence of illness or injury, such as sneezing, coughing, lameness or wounds. Behavioral metrics were responses to stranger-approach, used as a proxy indicator of socialization. These were organized into three categories: red, indicating a fearful response to approach, green, indicating an affiliative or neutral response, and yellow, indicating an ambivalent response. In kennels of ≤20 dogs, all were scored prior to and one year after implementing the standards; otherwise 20 bitches per kennel were randomly scored. Few dogs (<0.05%) had readily observed physical health problems both pre- and post-standards implementation. Improvement in the behavioral scores of dogs was seen at 83% of the sites that had implemented the standards for a full year (n=12). Development of the standards, now licensed and operating in the USA as a voluntary, third-party audited program, Canine Care Certified, will be discussed.

**A rotation designed to teach final-year veterinary students about dairy cattle welfare**

*Todd Duffield, Lena Levison and Derek Haley*
*University of Guelph, Population Medicine, Ontario Veterinary College, Department of Population Medicine, OVC, N1G 2W1, Canada; tduffiel@uoguelph.ca*

A one-week elective rotation for final year veterinary students was created at the Ontario Veterinary College (OVC) in 2016 to teach current scientific knowledge and practical elements of dairy cattle welfare. The rotation was designed with the following objectives in mind: 1. Review of existing dairy welfare standards and recognize contraventions, 2. Experience carrying out practical on-farm dairy cattle welfare assessments, 3. Practice communicating with clients and colleagues about animal welfare. The rotation is offered twice in an academic calendar year to a maximum of 12 student participants for each occurrence. The rotation was developed in part, as a response to a curriculum mapping exercise that identified gaps in welfare training in the veterinary curriculum and in part from a funding opportunity for dairy welfare training for veterinary students through Saputo. Six spaces in each rotation are held for OVC students, while students from other Canadian or American veterinary schools are encouraged to apply for the remaining placements. The week is structured as a balance of seminar style teaching, on-farm teaching, student exercises, and discussion around key and current dairy welfare issues. The week begins with a discussion around important dairy cattle welfare issues from the three perspectives of natural living, affective states, and biological function. This discussion leads to a brainstorming session around these important and current dairy welfare issues, and assignments to provide peer-to-peer summary reports on the welfare subjects. Other seminar components include topics such as cull cow decision-making, disbudding procedures, training around animal care assessments and ProAction, and management of downer cows and euthanasia decisions. Students are given practice with hock, neck and knee lesion scoring, lameness scoring, and body condition scoring. Visits include a tour of both a local veal farm and a livestock sales facility. For each rotation, students are divided into two groups and each group is assigned a dairy herd on which to conduct an animal care assessment using the ProAction requirements as a guide. Students prepare a report from their assessment, present their results orally to the class and the report is provided to the herd veterinarian to be shared with the herd owner. Students are evaluated through participation, presentations on key dairy welfare issues and farm reports prepared after conducting a proAction® animal care assessment. To date electives have been filled to capacity with students from all 5 veterinary colleges in Canada, Atlantic Veterinary College, Université de Monréal, Faculté de Médecine Vétérinaire, OVC, Western College of Veterinary Medicine and the University of Calgary Faculty of Veterinary Medicine. In addition, veterinary students from the University of Illinois and Michigan State University have also participated. The rotation effectiveness has been evaluated with student feedback and this has been extremely positive for the initial four offerings of the rotation over the past two years. The financial support of Saputo in helping create this elective experience is gratefully acknowledged.

**Perspectives of academic and industry alliance in education and research toward poultry health and welfare**

*Jean-Marc Larivière[1] and Daniel Venne[2]*
*[1]Institut de technologie agroalimentaire (ITA) Campus de La Pocatière, Ministère de l'Agriculture, des Pêcheries et de l'Alimentation du Québec (MAPAQ), 401, rue Poiré, G0R1Z0, La Pocatière, Québec, Canada, [2]Couvoir Scott, 1798 Rte Du Président-Kennedy, G0S3G0, Scott, Québec, Canada; jean-marc.lariviere@mapaq.gouv.qc.ca*

The Institut de technologie agroalimentaire (ITA), with campuses in La Pocatière and St-Hyacinthe, Québec, teaches future agricultural technicians at college and pre-university levels. ITA La Pocatière is the oldest agricultural school in Canada. Animal production technology and farm management technology programs cover poultry husbandry and welfare aspects for egg layers, broiler breeders, chickens, turkeys, waterfowls and game birds. As a priority, the first year's program emphasizes poultry welfare and familiarizes students with the Canadian Code of practice for the Care and Handling of Hatching Eggs, Breeders, Chickens and Turkeys (National Farm Animal Council). Partnerships with the industry (e.g. hatcheries and feed mills) as with universities, have been established for the benefit of education and research. Students are actively involved in experiments conducted at our farm and our partners share results of collaborative and financed research. Previous experiments have included effects of coccidiosis vaccines, weight rather than age in predicting ideal vaccination timing (e.g. infectious bursal disease), nutrition on muscle integrity and early detection of muscle damage (myopathy). Assessment of thermal comfort in chicks using infrared technology, gait score, litter quality and pododermatitis incidence in chickens have also been performed by students during experiments. Such studies have proposed practical solutions to relevant health and welfare problems and expose our students to realistic cases. Students are therefore trained in collaboration with industry partners (veterinarians, nutritionists, scientists and technicians), acquire a wealth of knowledge and develop valuable experience. A constant review of our practices is essential and will provide the poultry industry skilled technicians and farm managers capable of ensuring optimal animal health, welfare standards and sustainability.

## Impact of behavior assessment on animal maltreatment identification in a zoo prosecution case

*Vanessa Souza Soriano[1], Roberta Sommavilla[1], Sérvio Túlio Jacinto Reis[2] and Carla Forte Maiolino Molento[1]*
*[1]Federal University of Paraná, Animal Welfare Laboratory, Rua dos Funcionários 1540, 80035-050 Curitiba, Brazil, [2]Federal Police Department, Technical-Scientific Sector, Rua Profa Sandália Monzon 210, 82640-040 Curitiba, Brazil; carlamolento@ufpr.br*

There is increasing demand for action against animal cruelty. Animal welfare assessment as proposed by the Protocol for Expert Report on Animal Welfare (PERAW) may improve the identification of animal maltreatment in a variety of scenarios. Our objective was to describe the impact of behavior assessment within PERAW for an official report on a case of animal maltreatment suspicion in a Brazilian zoo, as formally requested by a public attorney in 2017. The PERAW was applied for 97 animals: one tiger, two lionesses, one dromedary, two camels, 41 peacocks, three greater rhea, 11 emus, 32 fallow deer and four red deer, all species cited in a legal prosecution against the zoo. The protocol includes four indicator groups: nutritional, comfort, health, and behavioural conditions, integrated into a single result regarding the existence of animal maltreatment. The analyses allowed for the identification of eight (8/97) maltreated animals: four red deer, both lionesses, the dromedary and the tiger. The group of behavioural conditions was significant in identifying the presence of maltreatment. For the dromedary, the maltreatment situation was detected solely on the basis of inadequate behavioural conditions, including social isolation, scarce grazing possibility and high frequency of fly avoidance behavior. The conclusion regarding red deer behavioural conditions was due to the impossibility for grazing, permanent exhibition of behaviors related to fear, a high respiration rate and searching for the farthest spot from the public in the small enclosure. In addition, one red deer exhibited pain behavior and another exhibited oral stereotypy. Behavioural conditions were considered inadequate for the lionesses because they were kept captive in the off-exhibit holding area for 16.5 out of 24 h, including the night period, characterizing extensive behavioural restriction. For the tiger, the conditions were inadequate because he was kept captive in the holding area, and within the feasible observation time exhibited extreme locomotion stereotypy (more than 50% of a focal 20 min-observation period) and no mitigation strategy was presented by the zoo. Additionally, indicators of comfort showed that conditions were inadequate for four red deer, both lionesses and the tiger; health indicators showed that the health of one red deer and one lioness were inadequate; and nutritional indicators showed that the nutrition of one red deer was inadequate. The impact of behavior assessment seems due to the possibility of acquiring information regarding animal conditions from their own perspective, collaborating to a better understanding of the welfare context. In conclusion, the assessment of behavioural conditions was decisive to the identification of animal maltreatment, showing the critical importance of behavior assessment as one of the groups of indicators to report on animal maltreatment.

## Female mice can distinguish between conspecifics raised with enrichment and those raised in standard 'shoebox' cages

*Aimée Marie Adcock, Emma Nip, Basma Nazal, Aileen Maclellan and Georgia Mason*
*University of Guelph, Animal Biosciences, 50 Stone Rd E, Guelph, ON N1G 2W1, Canada;*
*abegley@uoguelph.ca*

In mink and drosophila, the only two species studied, males raised with enrichment (E) win more copulations with females than do males raised without, likely reflecting females' evolved tendencies to assess stress when choosing their mates. But in social species, do females similarly prefer E females over non-enriched (NE)? We investigated this using female mice. Generally, NE females are more aggressive. They are also more stereotypic and likely to show depression-like changes (learned helplessness & anhedonia) – effects that human research suggests reduces both social attractiveness and sociability. We, therefore, tested the hypotheses that female mice can distinguish conspecifics raised in different housing types, and that NE-raised females are less attractive as companions and less interested in social contact. C57BL/6 mice (hereafter 'focal mice') from E (n=20) or NE (n=20) cages were exposed to two unfamiliar female stimulus mice from another strain (either DBA/2 or BALB/c), also from either E (n=20) or NE (n=20) cages, in a modified social preference apparatus (a square arena where stimulus mice are enclosed in opposite corners, and novel objects are enclosed in the other 2 corners). This was repeated for 10 min/day until focal mice were familiar with both stimulus mice [inferred from investigatory sniffing, which wanes with familiarity]). Trios were then filmed for 30 minutes and an observer blinded to treatment scored focal mouse proximity to each corner enclosure. Data were analysed using general linear mixed models (GLMMs). As predicted EE mice spent more time near other mice than did NE focal mice ($F_{(1,30.44)}$=13.75, $P$=0.0008). However, this reflected an interaction with stimulus mouse housing ($F_{1.32.59}$=6.74, $P$=0.01) caused by EE focal mice spending more time with NE stimulus mice ($F_{1,18}$=3.39, $P$=0.08), but NE focal mice spending more time with EE stimulus mice ($F_{1,15}$=4.84, $P$=0.04). Our hypothesis that female mice can discriminate E- from NE-raised conspecifics was thus supported. However, our hypothesis that NE-raised mice would be less attractive as companions was rejected (since this held only for NE focal mice), as was our hypothesis that they would be less social (since this held only for NE stimulus mice; $F_{1, 31.45}$=17.64, $P$=0.0002). To determine what underlies these complex effects, and identify which cues focal mice were using, we investigated the role of stimulus mouse homecage behaviour. Time spent inactive but awake (IBA, a depressive-seeming behaviour that covaries with learned helplessness) seems to play a role. For NE focal mice, the group that avoided NE stimulus mice, time spent near NE stimulus mice was negatively related to stimulus mouse IBA behaviour ($F_{(1,12.99)}$=6.15, $P$=0.03). This did not hold for EE focal mice. Research and analyses are still ongoing, but together these findings provide the first evidence that female mice can distinguish between NE and EE conspecifics; and cautiously suggest that NE mice avoid other NE mice who show depression-like changes.

## Enriched mice are nice: long-term effects of environmental enrichment on agonism in female C57BL/6s, DBA/2s, and BALB/cs

*Emma Nip, Aimee Adcock, Basma Nazal and Georgia Mason*
*University of Guelph, Animal Biosciences, 50 Stone Rd E, N1G 2W1, Guelph, ON, Canada;*
*emmanip@yahoo.co.uk*

Laboratory mice usually live in non-enriched (NE) shoebox-sized cages. Adding environmental enrichment (EE), preferred stimuli that provide complexity (often combined with increased space), generally enhances animal welfare, but in male mice, it can trigger resource-defence agonism. In contrast, it seems to reduce agonism in female mice, with one study linking this to reduced stereotypic behaviour (SB). Here, using females of 3 strains – C57BL/6 (C57s), DBA/2 (DBAs) and BALB/c (BALBs) – we sought to replicate these findings; see whether they change over time; assess whether EE affects agonism qualitatively, not just quantitatively; and investigate how such effects relate to both SB and the lower interaction rates likely to occur in larger cages. At weaning (between 3-4 weeks of age),165 females were placed into mixed strain trios: 99 mice in NE cages, and 66 in 60×60×30 cm EE cages containing tubes, wheels and chewable items. We scan-sampled in-cage behaviour for 4 h/day for 5-6 days, in four observation periods (2, 4, 7 and 9 months post-weaning). Agonism included aggression (mounting, rough grooming, chasing & pinning), displacement (pushing) and sniffing. Replicating previous work, EE mice consistently displayed less agonism (by c.15-25%) than NE mice ($F_{1,282.5}$=120, $P$<0.0001). In NE C57s, the most agonistic sub-group, levels of agonism remained high and did not change over time ($F_{3,110}$=0.73, $P$=0.54). However, these mice received less agonism over time ($F_{3,110}$=9.28, $P$<0.0001), while reciprocally, NE DBAs and BALBs displayed less agonism over time ($F_{3,126}$≥5.37, $P$≤0.0016), although never to the low levels seen in EE mice. In contrast, in all EE mice, agonism remained consistently low over time (NS effect of observation period), suggesting that they may form stable hierarchies faster than NE mice. Turning to qualitative aspects, a smaller proportion of agonism was aggressive in EE C57s ($F_{1,46}$=11.3, $P$=0.0016), though the opposite held for EE BALBs (the least agonistic sub-group) ($F_{1,52}$=12.6, $P$=0.0008); and EE C57s and DBAs also performed proportionately more investigatory sniffing: a relatively affiliative agonistic behaviour (respectively $F_{1,46}$=25, $P$<0.0001; $F_{1,47}$=5.19, $P$=0.027). Lastly, there were fewer social interactions overall in the large EE cages than the small NE cages ($F_{1,167.6}$=119.76, $P$<0.0001), but this did not explain the decreased agonism: in all save the last observation period, EE mice performed less agonism even after correcting for their decreased social interactions (2 months: $F_{1,52.4}$=6.71; 4 months: $F_{1,51.83}$=14.65; 7 months: $F_{1,50.5}$=13.12; $P$≤0.012 for all), suggesting that this effect is not just a by-product of having more space. We could not, however, replicate prior findings linking low SB with reduced agonism. Overall, providing enrichment reduces agonism in female mice, adding to its welfare benefits. Underlying mechanisms now need investigating.

### Early pup mortality in laboratory mouse breeding: presence of older pups critically affects perinatal survival

*Sophie Brajon[1], Gabriela Munhoz Morello[1], Jan Hultgren[2], Colin Gilbert[3] and I. Anna S. Olsson[1]*
*[1]i3S, Instituto de Investigação e Inovação em Saúde, Universidade do Porto, Laboratory Animal Science, Rua Alfredo Allen 208, 4200-135 Porto, Portugal, [2]Swedish University of Agricultural Sciences, Department of Animal Environment and Health, P.O. Box 234, 532 23 Skara, Sweden, [3]Babraham Institute, Veterinary Sciences Department, Babraham Research Campus, Cambridge, CB22 3AT, United Kingdom; sophie.brajon@i3s.up.pt*

Perinatal mortality is frequent in laboratory mouse breeding and is often manifested in the disappearance of pups overnight. The present research was designed to study how distinct conventionally used social breeding configurations affect pup survival. A total of 209 litters of C57BL/6 mice were studied across two licensed breeding facilities (109 in Breeding Facility 1,BF1, and 100 in BF2). Litters studied in BF1 were 54 from single-housed dams (S), 35 from mice housed in trios without any other litter present in the cage (T0) and 20 from mice housed in trios with an additional litter in the cage T1). Litters studied in BF2 were 41 from T0, and 59 from T1. Litters from both breeding facilities were checked daily to track pup mortality, while litters from BF1 were further followed through video-recordings for determining the incidence of cannibalism and infanticide. There was no experimental manipulation of housing or husbandry. Mortality means were compared across housing treatments through two-tailed t-tests, considering 95% confidence level. Mortality was at least twice as high when another litter was present both in BF1 (T0 33.2%, T1 60.3%, S 36.0%; $P<0.05$) and in BF2 (T0 16.8%, T1 59.9%; $P<0.01$). Moreover, the incidence of litters having 100% mortality (i.e. loss of all pups) was at least twice greater in T1 (50.0 and 37% of litters in BF1 and BF2, respectively) compared to T0 (22.9 and 10% of litters in BF1 and BF2, respectively). On the other hand, in BF2, 68% of the litters in T0 had zero mortality rate, whereas there were no litters with zero mortality in T1, although the same mortality pattern was not seen in BF1 (with 14.3 and 15.0% of zero mortality litters in T0 and T1, respectively). Video observations in BF1 revealed more pup deaths than initially detected during daily cage checking. Of the 401 deaths recorded, 166 pups were cannibalised and 73 of these were eaten before the first pup counting, thus only detected through the video observations. Cannibalism of dead pups was more often performed by the dam and female cage mate than by the male (98.4, 99.2 and 51.2% respectively). Active infanticide was only observed 8 times. The cause of death could in most cases not be determined neither from video recordings nor from post-mortem examination of remaining pups. These results show: (1) that the presence of older pups in the cage is a high risk factor for pup mortality; and (2) that the perinatal mortality in laboratory mouse husbandry is considerably underestimated since many dead pups are cannibalized before they are found at cage checking. Present housing and husbandry practice need reconsideration.

## The best things come in threes: assessing recommendations to minimize male mouse aggression

*Brianna N. Gaskill[1] and Paulin Jirkof[2]*
[1]*Purdue Univ., 270 S. Russell St, 47906 West Lafayette, USA, [2]Univ. of Zurich, Rämistrasse 71, 8006 Zürich, Switzerland; bgaskill@purdue.edu*

US guidelines recommend social housing for laboratory mice however inter-male aggression can often lead to severe injury and death, making long-term studies challenging. Previous literature recommends small group sizes (<5 mice) and establishing groups as early as possible, however the ideal group size <5 or the ideal age at grouping to reduce aggression have not been empirically tested. Additionally, increased aggression is observed after injections or cage cleaning. We hypothesized that: (1) smaller groups sizes; (2) earlier age of allocation to an experimental group/cage decreases aggression; and (3) manipulation of animals increases aggression in male mice. A 14 wk study was performed to assess the following treatments in male CD-1/ICR mice in a balanced factorial design: group size (1,2,or 3), age at grouping (5 or 7 wks of age), and manipulation (daily scruffing and weekly minimal handling). Data were averaged per cage (n=5; 60 cages/120 mice). All mice received 4 g of nesting material and had *ad libitum* food and water. Visual inspection for severe wounding was done daily. Body weights, food consumption, and nest scores were documented weekly and sucrose consumption, as a measure of anhedonia, was tested at 0, 1, 4, and 8 wks. At the end of the study mice were euthanized then pelted to assess wounding with the pelt aggression lesion scale (PALS). Fecal corticosterone and hematology data was collected but is not presented. Data were analyzed as a GLM. Group size and age at grouping affected body weight ($F_{2,755}$=42.2;$P<0.001$). Generally, mice in larger groups weighed more ($P<0.05$). However in groups of 3, mice combined at 7 wks were heavier on average than mice grouped at 5 wks ($P<0.05$). In mice where wounding could be visually observed, they were lighter than unwounded mice ($F_{1,755}$=14.1;$P=0.002$). Solitary mice, housed at 7 wks and scruffed weekly consumed the most food per mouse ($P<0.05$). Solitary mice grouped at 5 wks built higher scoring nests than did groups of 2 or 3 combined at the same age ($P<0.05$). Solitary mice ate significantly more sucrose per mouse than did cages of 2 or 3 mice ($P<0.05$). Very little fighting was observed during daily checks. However group size significantly affected the cage average PALS ($F_{2,47}$=4.3;$P=0.019$). Pair-housed mice had higher PALS and more overall wounding than groups of 3 ($P<0.05$). Minimally handled mice also had higher PALS than daily scruffed mice ($F_{1,47}$=4.3;$P=0.04$). Solitary housing is considered negative for social species. Therefore, it was surprising to see solitary mice consume more sucrose, suggesting they were less anhedonic or depressed. However, high nest scores, food consumption, and low body weights indicate solitary mice may be experiencing thermal stress, even with the provision of nesting material. Minimal handling increased wounding, another unexpected result. While age altered our measures, there was no strong effect on mouse welfare. Based on this data CD-1 mice can successfully be housed for up to 14 wks and groups of 3 may be the best for reducing even minor levels of aggression.

### Behavioural evidence of curiosity in zebrafish

*Becca Franks[1,2], Leigh Gaffney[1], Courtney Graham[1] and Daniel Weary[1]*
[1]*University of British Columbia, Animal Welfare Program/Faculty of Land and Food Systems, 2357 Main Mall, Vancouver BC, V6T 1Z4, Canada,* [2]*New York University, Animal Studies, 285 Mercer St, New York City, NY 10003, USA; beccafranks@gmail.com*

Curiosity, the motivation for information, has been found in species across the animal kingdom. Under free-choice paradigms and in contrast to neophilia (generic and fleeting attraction to novelty), curiosity can be a measure of good welfare. Moreover, as curiosity indicates the capacity to want cognitive stimulation, evidence of curiosity in a species suggests the need to consider cognitive enrichment programs. As yet, however, the concept of curiosity has not been explicitly studied or distinguished from simple neophilia in fish. To determine whether fish show behavioral evidence of curiosity, we presented 30 different novel objects to zebrafish held in semi-natural conditions and in accordance with the ISAE ethical guidelines. Following various theories of curiosity, objects were rated by human coders along three dimensions (complexity, ecological relevance, and predator resemblance) and presented to the fish one-at-a-time in a randomized order over the course of a month (6 tanks; 10 fish/tank; 1-2 objects/day, 10+ min. between presentations). We hypothesized that the following behaviors would constitute evidence for curiosity: (1) swift behavioral approach; (2) sustained interest; and (3) differential interest by object. We coded visual inspections during the first and last 100 seconds of the 10-minute presentation period (start and end periods, respectively) and social behavior during those periods in addition to a baseline period one hour before object introduction. Multilevel models of inspection behavior (crossed-random effects to control for repeated observations of tanks and objects) revealed evidence of curiosity. Zebrafish readily approached all objects (1 s median latency) and showed signs of sustained attention to some: within the start period, those objects attracting the most attention in the first 50 s continued to attract higher attention in the second 50 s ($\rho=0.57$; $P<0.001$). For example, a white shell attracted 6.8±1.8 (mean ± standard error) inspections during the first 50 s and 4.2±1.1 inspections during the second 50 s, compared to 1.9±0.3 and 1.5±0.3 inspections for a blue toy van. While we found consistency across tanks in terms of which objects attracted the most attention (likelihood ratio test: $\chi^2(1)=20.22$, $P<0.001$), the only object dimension to predict sustained interest was (low) predator resemblance ($\rho=0.34$, $P<0.01$). Compared to baseline, exposure to the objects altered social dynamics during the start period: decreasing aggression ($t(4.81)=6.87$, $P<0.01$) and increasing group cohesion ($t(5.23)=4.37$, $P<0.01$) and coordination ($t(5.11)=6.97$, $P<0.01$). The presence of curiosity in zebrafish points towards its deep evolutionary roots and could bear implications for fish welfare. These results suggest that, like other animals, zebrafish may be motivated to seek out cognitive stimulation – a motivation that may need to be satisfied for them to live a good life.

**Routine water changes in enriched environments temporarily increase stress and alter social behavior in zebrafish**

*Christine Powell[1], Marina A.G. Von Keyserlingk[1] and Becca Franks[2]*
*[1]University of British Columbia, Animal Welfare Program, 2357 Main Mall, Vancouver BC, V6T 1Z4, Canada, [2]New York University, Animal Studies, Department of Environmental Studies, 285 Mercer Street, New York City, NY 10003, USA; christinempowell@live.com*

Fish have recently become one of the most used animals in scientific research. The current scientific consensus is that fish have the capacity to suffer, emphasizing the importance of fish welfare research. Environmental enrichment can increase laboratory fish welfare, but it may come with unintended welfare costs: enriched aquatic environments require different husbandry procedures than standard laboratory housing (e.g. manual water changes) that could elevate stress. Previous research with zebrafish has examined husbandry procedures in non-enriched environments and has focused on physiological indicators of health; no research has yet investigated the effects of husbandry on social behavior, which for a highly social species such as zebrafish, may be a sensitive indicator of welfare. We predicted that the social behavior of zebrafish would be negatively impacted by manual water changes. We tested our prediction by measuring three social behaviors – agonism, group cohesion, and coordinated swimming – as well as the location of the fish in the water column as bottom-dwelling is considered an avoidance behavior indicative of negative states in zebrafish. Routine weekly water changes consisted of manual scrubbing, siphoning, and changing water depth as soiled water was removed and replaced in a process that averaged 15 minutes. Over the course of 4 weeks, we video recorded 5 tanks (10 fish per 110 l tanks) on 8 days: 2 before-cleaning days (which served as our baseline), 4 cleaning days, and 2 after-cleaning days. From these videos, we coded social behavior and fish locations in the water column during two time periods after the husbandry event (time-matched on the non-cleaning days): (1) when the technicians were still in the room and (2) approximately one hour after they left the room. This sampling procedure produced 80 observations per behavior (5 tanks × 8 days × 2 time-periods) that we analyzed with generalized multilevel models, controlling for repeated sampling of tanks with a random intercept model. Directly after the water change event fish were less cohesive ($P<0.01$), less coordinated ($P<0.01$), and tended to be more aggressive ($P<0.07$); fish also spent more time lower in the water column, a sign of anxious avoidance ($P<0.01$). An hour after the technicians had left, aggression remained elevated ($P<0.04$), but all other behaviors returned to baseline ($P>0.1$). On the day after cleaning, most behaviors during both time periods were indistinguishable from baseline ($P>0.1$), except aggression, which was higher while the technicians were in the room ($P<0.04$) and remained marginally higher after they had left the room ($P<0.08$).

**Individual variability in motivation to access a food reward**

*Anna Ratuski, Lucia Amendola and Daniel Weary*
*University of British Columbia, Animal Welfare Program, 2357 Main Mall, Vancouver, BC, V6T*
*1Z4, Canada; aratuski@alumni.ubc.ca*

Motivational tests are typically based on access to a reward, often treating food rewards as the gold standard, but there has been little research on how individuals vary in their motivation to access these rewards. The aim of this study was to measure rats' individual motivation for a food reward (Cheerios). Rats (n=11) were tested in a cage with rewards hidden under a layer of sand. Twenty-one Cheerios were hidden in predetermined arrangements, with the dispersal of rewards systematically increasing in consecutive trials. Rats had one trial to become familiar with the task and were then tested three times. Each trial ended when the rat abandoned the test cage. Time spent searching and number of rewards found per trial were analyzed using Linear Mixed Models with rat identity as a random intercept. Rat identity explained 57% of the variation in searching duration and 60% of the variation in number of rewards found (n=11). Mean searching duration varied among rats from 38 s to 286 s (mean± SD of 175±26 s). Mean number of rewards found ranged between 6 and 20 Cheerios, with a mean of 14.2±1.6. Searching time and number of rewards found were both repeatable within rats (repeatability: r=0.59, $P<0.001$, and r=0.61, $P<0.001$, respectively). These results indicate that rats vary considerably in their individual motivation to access a food reward, potentially influencing the results of studies based upon this motivation.

**Assessing stress in rabbits during handling: using infrared thermography to measure peripheral temperature fluctuations**

*Mariana Almeida[1], Vitor Pinheiro[1,2], Sandra Oliveira[3], George Stilwell[4] and Severiano Silva[1,2]*
*[1]CECAV, Universidade de Trás-os-Montes e Alto Douro, P.O. Box 1013, 5001-801 Vila Real, Portugal, [2]Departamento de Zootecnia, Universidade de Trás-os-Montes e Alto Douro, P.O. Box 1013, 5001-801 Vila Real, Portugal, [3]UTAD, Universidade de Trás-os-Montes e Alto Douro, P.O. Box 1013, 5001-801 Vila Real, Portugal, [4]CIISA, Faculty of Veterinary Medicine, University of Lisbon, Alto da Ajuda, 1300-477 Lisboa, Portugal; mdantas@utad.pt*

Handling is a necessary part of commercial rabbit production and is known to cause fear and stress in rabbits. Also, inappropriate handling of rabbits can cause lesions and lead to health problems. Therefore, identifying the most suitable handling procedure is a critical aspect of maintaining optimal rabbit welfare. It has also been shown that stress causes changes in core body temperature as well as in eyes, muzzle and ears. Infrared thermography (IRT) technology has been widely used in studies relating to health and welfare of farm animals, since it allows a remote reading of temperature, with no physical contact with the animal. With IRT it is possible to identify changes in peripheral temperature which results from changes in blood flow as a stress-induced response. This study aimed to assess the stress of rabbits during two different types of handling procedures, using ear, eye and muzzle IRT temperature fluctuations. The research was carried out with 50 hybrid male rabbits (New Zealand × California). The animals were taken from the home cages to be weighed, and two types of handling were applied: (1) 25 animals were grabbed by the back of the neck with an arm along the ventral abdomen; (2) the other 25 were grabbed gently with two hands placed on the abdomen. An infrared camera FLIR F4 was used to capture the thermal images of the eye, ear and muzzle. Temperature data was extracted from thermal images using the FLIR Tools. The thermal image readings took place at two stages: before and after handling. The measurements of IRT temperatures in the eye, ear and muzzle were measured at three separate sessions. Data were analyzed using repeated measures two-way ANOVA following least significant difference Student's t-test as comparisons test. All statistical analyses were performed using the JMP-SAS software. The results showed significant differences between stages of handling on the eye ($P<0.001$), ear ($P<0.021$) and muzzle ($P<0.001$) temperature. For all the thermal image reading sites, the lower temperature values ($P<0.05$) were observed after the handling procedure. Eye, ear and muzzle temperature decreased 2.72% ($35.94\pm0.07$ to $34.99\pm0.09$ °C); 2.74% ($29.24\pm0.17$ to $28.46\pm0.29$ °C) and 6.20% ($26.03\pm0.08$ to $24.51\pm0.10$ °C), respectively, after the procedure. Regarding the handling procedure, no effect was observed which means the two types of handling chosen for this study seem to cause the same level of stress to rabbits. IRT was able to detect small decreases in the eye, ear and muzzle temperature which is probably due to vasoconstriction after catecholamine release. Further research is needed to correlate IRT with other stress indicators to ascertain the value of this technique.

**Seek and hide: understanding pre-partum behavior of cattle by use of inter-species comparison**

*Maria Vilain Rørvang[1,2], Birte L. Nielsen[3,4], Mette S. Herskin[2] and Margit Bak Jensen[2]*
*[1]SLU, Dept. of Biosystems and Technology, Sundsvägen 16, 23053 Alnarp, Sweden, [2]Aarhus University, Dept. of Animal Science, Blichers Allé 20, 8830 Tjele, Denmark, [3]INRA, Modélisation Systémique Appliquée aux Ruminants, Université Paris-Saclay, 75231 Paris, France, [4]INRA, NeuroBiologie de l´Olfaction, Université Paris-Saclay, 78350 Jouy-en-Josas, France; mariav.rorvang@slu.se*

The event of giving birth is an essential part of animal production. In dairy cattle production, substantial economical and welfare-related challenges arise around the time of calving, and hence focus is placed on efficient management of the parturient cow. Aiming to understand the biological basis of bovine pre-partum maternal behavior, this literature review, based on studies of managed and feral cattle breeds, examines similarities and dissimilarities to other members of the ungulate clade. It is clear from the literature that pre-partum maternal behavior varies among species; however, the final goal of ungulate mothers is the same: ensuring an optimal environment for a calm parturition and onset of post-partum maternal behavior by locating an appropriate birth site with low risk of predation, disturbances, and mistaken identity of offspring. Specific features of chosen birth sites vary among species, and largely depend on characteristics of the environment, e.g. level of vegetative cover, and ungulate mothers display a considerable ability to adapt to their surroundings. Within commercial dairy housing, however, the cows' ability to adapt appears to be challenged. The indoor production environment leaves little room to express birth-site selection behavior due to e.g. high stocking density. This poses a risk of agonistic social behavior, disturbances, and mis-mothering, as well as exposure to olfactory cues influencing pre- and post-partum maternal behavior. Dairy cows are exposed to several factors in a commercial calving environment, which may thwart their pre-partum motivations and influence their behavior accordingly. In addition, pre-partum cows may be more affected by olfactory cues than other ungulate females (e.g. sheep) because cows are already attracted to birth fluids before calving. At present, the motivations underlying the apparent pre-partum isolation seeking behavior of cows have not been fully explored. Additionally, concepts traditionally applied to ungulate maternal behavior, such as the hider/follower-dichotomy, appear overly simplistic. Based on the reviewed literature, we suggest that more scientific focus be given to pre-partum maternal behavior (i.e. the phase of birth-site selection) of dairy cows, as many factors influence their behavior – and potentially their welfare – at this point. Providing dairy cows with an environment where they are able to perform pre-partum maternal behavior may help to ensure a calm and secure calving and provide optimal surroundings for post-partum maternal behavior. Future research focusing on housing systems adapted for the motivations of parturient cows is needed to determine the importance of degree of visual cover and distance from the group within the constraints of indoor dairy housing systems. Ultimately, this knowledge may be advantageous for future development of housing and management systems for dairy cows and calves.

**Effect of parity on dairy cows and heifers' preferences for calving location**

*Erika M. Edwards[1], Katy L. Proudfoot[2], Heather M. Dann[3], Liesel G. Schneider[1] and Peter D. Krawczel[1]*
[1]*The University of Tennessee, Department of Animal Science, 2506 River Dr., Knoxville, TN 37922, USA,* [2]*The Ohio State University, Department of Veterinary Preventive Medicine, 1920 Coffey Rd., Columbus, Ohio 43210, USA,* [3]*The William H. Miner Agricultural Research Institute, 1034 Miner Farm Rd., Chazy, NY 12921, USA; eedwar24@vols.utk.edu*

Previous research indicates that dairy cows will express some of the same behaviors within confinement systems as they would on pasture at calving, including seeking an isolated area to give birth. Yet, it is still unknown what factors may drive preferences for a secluded calving environment. The objective was to determine the effects of parity, season and group size on dairy cows and heifers' preference for calving location when group housed and provided free access to pasture. This study was implemented using an observational study design. Multiparous Holstein cows (n=33) and nulliparous Holstein heifers (n=32) were dynamically enrolled -21 d prior to their expected calving date and removed on the day of calving. Cows had continuous access to a bedded-pack barn (area 1; 167.4 $m^2$), open pasture (area 2; 1.82 hectares), and an area of natural forage cover (i.e. trees and tall grasses; area 3; 0.24 hectares). Video data were used to determine calving location. Cows calved from August to December 2016. Season was categorized into summer (August 6 – September 21), fall (September 22 – December 20) and winter (December 21 – December 28). The social composition of the group was dynamic as cows entered and left the pasture based on expected and actual calving date. For each cow, mean daily group size was calculated over the days she was enrolled, including at calving. There were three possible outcomes for area of calving, making the dependent variable multinomial. Therefore, multinomial logistic regression was used to test if factors were associated with the probability of calving in the three areas of the environment, and model selection was performed by backward manual elimination. Twenty-five calvings (38%) occurred in the barn, 17 calvings (26%) occurred on open pasture, and 23 calvings (35%) occurred in the area of natural forage cover. Parity was associated with the location of calving ($P=0.02$); heifers were more likely to calve in the natural forage covered area compared to the barn (OR=5.88; 95% CI=1.69, 20.42). While, cows were less likely to calve in the natural forage cover area compared to the barn (OR=0.17; 95% CI=0.05, 0.59). However, heifers and cows were equally as likely to calve in open pasture or the barn (OR=2.89; 95% CI=0.8, 10.53 and OR=0.35; 95% CI=0.095, 1.26, respectively). Season, group size at calving, and mean daily group size were not associated with calving location preference ($P>0.25$). These results suggest parity plays a role in preference for calving location when group housed. Dairy producers are encouraged to consider these preferences when they design maternity areas for their heifers and cows.

## Use of qualitative behaviour assessment to evaluate styles of maternal protective behaviour in Girolando cows

*Maria C. Ceballos[1], Karen C. Rocha Góis[1,2], Françoise Wemelsfelder[3], Pedro H.E. Trindade[1], Aline C. Sant'anna[1,4] and Mateus J.R. Paranhos Da Costa[1,5]*
[1]Grupo de Estudos e Pesquisas em Etologia e Ecologia Animal (ETCO), V. Prof. Paulo Don. Cast S/N, 14884-900, Brazil, [2]Pós-Graduação em Zootecnia, FCAV-UNESP, V. Prof. Paulo Don. Cast S/N, 14884-900, Brazil, [3]SRUC, Roslin Institute Building, Easter Bush, EH25 9RG, United Kingdom, [4]Dpto de Zoologia, ICB, UFJF, R. José Lourenço Kelmer, S/N, 36.036- 330, Brazil, [5]Dpto de Zootecnia, FCAV-UNESP, V. Prof. Paulo Don. Cast S/N, 14884-900, Brazil; mceballos30@gmail.com

An important aspect of cow temperament is maternal protective behaviour. Aggressive cows tend to be more agitated during handling, increasing the risk of injury to handlers, herself and her offspring. The aim of this study was to test the use of Qualitative Behaviour Assessment (QBA) in evaluating different styles of maternal protective behaviour in dairy cows. The study was based on 20 videos of Girolando cows, recorded just after calving during first handling of their calves, when calves were physically restrained for identification and navel care in a commercial dairy farm in Minas Gerais State, Brazil. Eight students from a post-graduate program in Animal Science were asked to score the cows' behavioural expressions observed in each video on a visual analogue scale (125 mm length), using 13 predefined adjectives (relaxed, worried, angry, attentive, frightened, permissive, indifferent, lovely, agitated, calm, aggressive, active, comfortable). Before starting the observations, the observers had two training sessions with 20 different videos. After the training sessions, each observer scored each video three times with a 15-day interval between sessions. Principal component analysis (correlation matrix, no rotation) was used to analyse all data together for the 8 observers across the 3 sessions. Inter-observer reliability was calculated for each session separately, using Kendall's coefficients of concordance (W) to correlate the 8 observers' scores for the 20 videos on the first three principal components (PC) in each session. Intra-observer reliability was calculated for each observer separately, using Kendall's W to correlate each observer's scores for the 20 videos on the first three PC across the three sessions. PC1 explained 33.2% of the total variance in the data set, characterizing the cows' response to handling as ranging from 'relaxed/comfortable' to 'worried/agitated'. PC2 (16.6%) ranged from 'indifferent' to 'attentive', while PC3 (11.5%) ranged from 'permissive' to 'aggressive'. Inter-observer reliability in the three sessions was strong for PC1 (W=0.70 to 0.73) and PC2 (W=0.71-0.77) scores, but moderate for PC3 scores (W=0.54-0.57). Intra-observer agreement for the 8 observers was strong for PC1 (W=0.71-0.89) and PC2 (W=0.74-0.90) scores. For PC3 scores agreement across sessions was strong for 6 observers (W=0.71-0.84), but moderate for 2 observers (W=0.63-0.65). These results indicate that QBA can discriminate in considerable detail between different types of responses by cows to the first handling of their calves, revealing substantial inter- and intra-observer reliability. QBA appears to be a promising method for characterizing maternal protective styles in Girolando cows.

## The effect of concrete feedpad feeding periods on oestrous behaviour in dairy cows in a pasture-based farming system

*Andrew Fisher, Caroline Van Oostveen, Melanie Conley and Ellen Jongman*
*The University of Melbourne, Animal Welfare Science Centre and Faculty of Veterinary and Agricultural Sciences, 250 Princes Highway, Werribee, Victoria 3030, Australia; adfisher@unimelb.edu.au*

The detection of oestrous behaviour is used to identify the optimal timing for artificial insemination in dairy cows, and can be particularly challenging in extensively managed dairy cows. Because of changing climatic conditions, more Australian dairy farmers have installed concrete 'feedpads' to enable the effective supplementation of grazed pasture, which may be limited in dry weather. This study was conducted to identify whether the use of feeding periods on these concrete feedpads affected the expression of oestrous behaviour in dairy cows. Following approval by the Institutional Animal Ethics Committee, 64 Holstein-Friesian cows (4.8 years (SD 1.51); 559 kg (SD 54.1); 84 DIM (SD 22.1)), were used. Feedpad cows (n=32) were placed on a feedpad for 78.9 (SD 0.34) min after each of morning and afternoon milking. Pasture-only cows (n=32) were allowed to walk straight back to the paddock after each milking. The diets offered were isoenergetic in that feedpad cows received 12 kg DM of a mixed ration on the feedpad daily, whereas pasture-only cows received 8 kg grain during milking and 4 kg silage while at pasture. Both cow groups received the same daily pasture allocation in adjoining plots. Continuous behavioural observations using two observers were conducted for approximately 9 h per day over a total of 6.5 days across paddock, dairy yard, feedpad and laneway environments, following an Ovsynch program. Behaviours recorded included flehmen, chin-resting, mounting attempts and mounting by instigator animals, and being mounted and standing heat in recipient animals. Statistical analysis excluded dairy and laneway data as there was little behaviour recorded. Data were divided into AM and PM periods, and Poisson regression was used to compare the ratio of incidence rates of behavioural categories, where Incidence Rate = # events of specific behaviour / # cow hours of observation in a specific location × 100. Data presented are the incidence rates. Feedpad cows showed less mounting and standing oestrous behaviour on pasture compared with pasture-only cows (Mounting behaviour AM: 42.9 vs 60.8; IRR=0.71; $P<0.006$; PM: 38.4 vs 67.5; IRR=0.57; $P<0.001$). Feedpad cows also showed less mounting behaviour on the feedpad compared with when they were at pasture (AM: 4.8 vs 38.4; IRR=0.13; $P<0.001$; PM: 13.2 vs 42.9; IRR=0.31; $P<0.001$). There were no differences between AM and PM for standing oestrous and mounting behaviour in either cow group. In conclusion, cows in a pasture system that includes feedpad feeding may need additional observation or detection aids for effective identification of oestrus for AI. Observations conducted on the feedpad, while convenient, are unlikely to prove beneficial.

**Effects of three types of bedding surface on dairy cattle behavior, preference, and hygiene**

*Karin Schütz[1], Vanessa Cave[1], Neil Cox[1], Frankie Huddart[1] and Cassandra Tucker[2]*
*[1]AgResearch Ltd, Ruakura Research Centre, Hamilton 3214, New Zealand, [2]University of California, Center for Animal Welfare, Department of Animal Science, 1 Shields Avenue, Davis, CA 95616, USA; karin.schutz@agresearch.co.nz*

Muddy surfaces have negative effects on the health and welfare of dairy cattle, and if possible, cows will avoid mud. However, it is unclear if it is the moisture content or the contamination with manure that is aversive to cows. This study aimed to assess the use and preference for different bedding surfaces: (1) clean and dry (CLEAN, dry matter content, DM: 44±2.8%); (2) dirty (DIRTY, contaminated with manure, DM: 40±3.7%); and (3) wet (WET, wetted by water, DM: 23±3.3%). All surfaces were 0.4 m deep bedded wood chip. Non-lactating, pregnant cows were tested individually (n=18, mean 24-h temperature: 9.9±4.46 °C, mean ± SD for all preceding values). Cows were kept for 18 h on wood chip and 6 h on pasture to allow for daily feed intake; no feed was provided in the test pens. To ensure cows made informed choices and to measure changes in behavior and hygiene associated with each option, they were first exposed to each surface for 5 days (n=12 per surface). Data were analysed using REML and t-tests. Cows on the wet surface spent the least amount of time lying (WET: 21%, DIRTY: 57%, CLEAN: 64%/18 h, $P \leq 0.005$) and spent more time lying when on pasture (WET: 13%, DIRTY: 4%, CLEAN: 3%/6 h, $P \leq 0.005$). The total lying times during the 5-day surface exposure were for WET: 4.6±0.80 h, DIRTY: 10.6±0.27 h, and CLEAN: 11.7±0.27 h/24 h, ($P \leq 0.006$). Cows on the wet bedding had fewer bouts of lateral lying (WET: 0.9±1.36#, DIRTY: 6.3±1.36#, CLEAN: 8.4±1.38#/18 h, $P=0.002$), spent less time lying with their heads supported (WET: 18.9±7.17 min, DIRTY: 36.7±7.17 min, CLEAN: 39.1±7.26 min/18 h, $P=0.047$), and spent less time with the front legs tucked (WET: 16±4.3%, DIRTY: 41±4.3%, CLEAN: 50±4.3% of time spent lying, mean ± SEM for all preceding values, $P<0.001$), than cows on the other surfaces. Cows on the dirty bedding were less clean compared to the other treatment groups (0.6 of a score on a 5-point scale, SED: 0.11 for both comparisons, $P \leq 0.021$). They were then given a free choice between 2 known surfaces for 2 consecutive days (n=6/pairwise comparison, n=18 total). Cows ranked bedding as CLEAN>DIRTY>WET ($P<0.001$). In summary, there is compelling evidence that wet bedding impairs the welfare of dairy cattle by affecting the quantity and quality of rest. Rebound responses after the exposure indicate that the motivation to rest is not fulfilled on wet surfaces. Finally, when given a choice, they show clear preferences that they will avoid wet and dirty bedding, indicating that changes in affective state likely underlie their behavioral responses.

## The impact of housing tie-stall-housed dairy cows in deep-bedded loose-pens during the dry period on lying time and postures

*Elise Shepley[1], Giovanni Obinu[2] and Elsa Vasseur[1]*
*[1]McGill University, 21111 Lakeshore Rd., Ste-Anne-de-Bellevue, Quebec, H9X 3V9, Canada,*
*[2]Università degli Studi di Sassari, 21 Piazza Università, 07100 Sassari, Italy; eshepley1@gmail.com*

A cow's housing environment can significantly impact her overall welfare, particularly during the dry period as the cow undergoes a number of managerial and physiological changes. As tie-stall-housed cows are normally housed in tie-stalls during both their lactation and dry period, an investigation into alternative housing options that could fulfill the requirements of prepartum cows is needed. Lying behaviors of dairy cows can be useful measurements for determining the level of comfort and opportunity of movement provided in these alternative housing environments. The objective of the current project was to determine if lying time and lying postures differ between dairy cows housed in a tie-stall vs a loose-pen during the 8-week dry period. Fourteen cows were paired based on parity and calving date and assigned at dry-off to either a deep-bedded straw loose-pen (LP) or a tie-stall with a rubber mat base and 2 cm of wood shavings (TS). Lying time was measured by leg-mounted pedometers. Twenty-four-hour video recordings were taken weekly for each cow by overhead-mounted cameras and the cows' lying postures were recorded by a trained observer using 1-min scan sampling of the video images for videos taken on the first, middle, and last week of the dry period. All data were analyzed using a mixed model with treatment, week, pair, and treatment × week interaction as fixed effects and cow nested within treatment and pair as a random effect. Daily lying time was numerically, but not significantly, higher for LP cows compared to TS cows (849.97±32.43 and 753.70±19.09 min/day, respectively, *P*=0.17). LP cows rested their head on their back more often than TS cows (8.58 vs 5.72% of the observed time, *P*<0.05). Similarly, LP cows exhibited greater variability in positioning of hind legs, keeping legs tucked in less often (73.19 vs 92.87%, *P*<0.01) in favor of alternative positions. Cows in LP also changed head position (22.02 vs 14.65 changes/24-hr recording for LP and TS, respectively, *P*=0.02), front leg position (22.24 vs 7.57, *P*<0.01), and hind leg position (88.75 vs 23.60, *P*<0.01) more often. Similarities in daily lying time between the two housing options suggest that neither environment impeded on the cows' lying abilities. The increased variety of postures and frequency of posture changes displayed by LP cows when lying may suggest that the cow is able to move with greater ease when provided with a combination of more space and a different lying surface, potentially increasing overall comfort for the cow. Lying behaviors as a means to evaluate the response of cows to different environments allow us to better understand the benefits of increasing movement opportunity for cows and the viability of alternative housing options during the dry period.

**Cross-contextual vocalisation rates and heart rates of Holstein-Friesian dairy heifers**

*Alexandra Green[1], Sabrina Lomax[1], Emi Tanaka[2], Ian Johnston[3] and Cameron Clark[1]*
*[1]University of Sydney, Dairy Science Group, School of Life and Environmental Sciences, Camden, NSW, 2570, Australia, [2]University of Sydney, School of Mathematics and Statistics, Camperdown, NSW, 2006, Australia, [3]University of Sydney, School of Psychology, School of Life and Environmental Sciences, Camperdown, NSW, 2006, Australia; a.green@sydney.edu.au*

Vocalisation rates and types are valuable indicators of emotional arousal and given their ease of capture, they can be used to assess on-farm welfare. In a range of mammalian species, increased arousal has been linked to an increase in vocalisation rate; with the expression of arousal differing between basic call types. Vocalisation rate has been studied in cattle experiencing oestrus, calf-separation and social isolation, yet a comparison of vocalisation rate in different emotionally arousing situations is yet to be conducted. The purpose of this study was to compare dairy cattle vocalisation rates of the two high frequency and low frequency vocalisation types in farming contexts that differ in emotional arousal. 18 Holstein-Friesian heifers of a similar age (24.5±2.5 months) were subject to different farming contexts: (1) a control situation in their paddock where they were left undisturbed; (2) housed in cattle yards to ensure both partial and full isolation from their conspecifics; (3) anticipation of a feed reward; and (4) a feed frustration situation in a paddock, where their feed access was denied. Vocalisation rate (calls/minute) and heart rate (bpm) were recorded in each of these contexts over five minutes and vocalisations were spectrographically analysed to categorise the vocalisation types. Based on the heart rate data (mean ± SEM), three different levels of arousal were apparent in the farming contexts ($P<0.001$). Heifers had the lowest heart rate in the control context (52.42±0.57) indicating the lowest level of arousal. The feed anticipation and frustration contexts elicited a medium arousal level, as per the heart rates of 57.49±0.83 and 56.80±0.77, respectively. The partial and full isolation contexts elicited the highest arousal in the heifers, as per the heart rates of 65.95±1.69 and 63.63±1.22, respectively. Contrary to the literature, high arousal contexts were associated with fewer ($P<0.001$) high frequency and low frequency vocalisations compared to the medium arousal contexts. Only 10 of the 18 heifers vocalised during high arousal compared to all heifers during medium arousal and none during the low arousal control. Within-individual vocalisation rates of both call types were stable in medium arousal (high vocalisations: r=0.17, P=0.002, low vocalisations: r=0.26, P=0.008) and high arousal (high vocalisations: r=0.48, P<0.001, low vocalisations: r=0.36, P<0.001). However, within-individual vocalisation rates differed when comparing medium arousal to high arousal (high vocalisations: r=0.09, low vocalisations: r=0.001). We conclude that rates of high frequency and low frequency vocalisations are both arousal-dependent and related to the individuality of the heifer. To further differentiate the vocalisations across contexts, our research will now determine the spectral and temporal components of the vocalisations, extracting vocal correlates of both arousal and valence.

**For the love of learning: heifers' motivation to participate in a learning task**

*Rebecca K. Meagher[1,2], Marina A.G. Von Keyserlingk[2] and Daniel M. Weary[2]*
[1]*University of Reading, School of Agriculture, Policy & Development, Whiteknights, P.O. Box 237, RG6 6AR, United Kingdom,* [2]*University of British Columbia, Animal Welfare Program, 2357 Main Mall, Vancouver, V6T 1Z4, Canada; rkmeagher@gmail.com*

Cognitive challenges appropriate to the species and individual have been suggested as a means for improving captive animal welfare, but the success of this approach has been little studied, particularly in farm animals. To test whether cattle are motivated to have learning opportunities, we trained 20 Holstein heifers to perform an operant response, a nose-touch to a rope, using a variable interval reinforcement schedule. The 'reward' was access to a small section of the training arena behind a gate, where they either received an opportunity to participate in a discrimination task (Learning) or received a matched feed reward without making a choice (Control). Heifers were assigned alternately to treatments based on speed of learning the initial nose-touch response. The discrimination task involved a choice of bins with lids of different colours (red or white) and textures, one of which contained a preferred feed (concentrate) and the other a less preferred feed (hay). Each heifer was trained daily with five trials per training session in the discrimination phase, until the Learning heifer performed at ≥80% accuracy for three consecutive days. The number of nose-touches and latency to approach the rope when allowed into the training area were used as measures of motivation to participate in the learning task. The Learning heifers had shorter latencies to approach than did Controls during the discrimination training (back-transformed means [95% CIs]: 17 [7-43] vs 67 [28-161] s; t-test, $P=0.026$). The difference in the number of rope touches was not statistically significant (Learning: 12.9±1.9 vs Control: 8.6±2.2; t-test, $P>0.10$). The observed patterns generally support the hypothesis that heifers were motivated to participate in the training; the lack of a significant difference in rope touches and the willingness of heifers in both groups to enter the training arena indicate that some aspects of the experience other than the discrimination learning itself may also have been rewarding for these heifers. Opportunities for learning therefore have potential to improve welfare in cattle, but it is not yet clear whether the cognitive challenge itself is necessary to achieve this goal.

## The effect of feeding and social enrichment during the milk-feeding stage on cognition of dairy calves

*Kelsey C. Horvath and Emily K. Miller-Cushon*
*University of Florida, Department of Animal Sciences, 2250 Shealy Drive, Gainesville, FL 32611,*
*USA; kchorvath@ufl.edu*

Environmental complexity affects cognitive development, but dairy calves are commonly housed in environments that restrict the performance of natural behaviours. We hypothesized that both nutritional and social enrichment would improve behavioral flexibility, as assessed in a cognitive task. Holstein heifer calves were randomly assigned at birth to 1 of 4 housing treatments differing in social environment and solid feed variety: individual housing without hay (IC; n=8) or with hay (IH; n=8), or group housing (4 calves/pen) without hay (GC; n=5) or with hay (GH; n=6). Milk was provided at 8 l/d via teat twice daily for individually housed calves and through an automated milk feeder for group housed calves, such that social housing treatments also differed in milk feeding method. All calves had free access to grain concentrate. At 5 weeks of age, calves were tested in a T-maze with a reward (0.2 l milk). After learning an initial reward location, calves were assessed in a reversal learning task to assess behavioral flexibility, where the reward location was changed to the opposite arm. Calves received five sessions/d for 4 d until they met a passing criteria (moving directly to correct side in 3 consecutive sessions) or reached a maximum of 20 sessions. We recorded completion time for individual sessions and the total number of sessions required to pass. Data were analyzed in a generalized linear mixed model, with session as a repeated measure for session completion time. The rate of passing was compared between treatments using Fisher's exact test. Of all calves, 75% (n=6) of IC, 80% (n=4) of GC, and 100% of IH and GH were able to learn the initial reward location and were tested in the reversal learning task. Enrichment affected pass rate for reversal learning (P=0.02), with more group-housed calves passing and IC having the lowest pass rate: IC: 33% (n=2), IH: 50% (n=4), GC: 100% (n=4), and GH: 100% (n=5). Both availability of hay (P=0.01) and social housing (P=0.004) reduced the sessions required to pass the reversal learning task (IC: 16.67, IH: 14.88, GC: 13.0, GH: 6.83 sessions; SE=2.2), and there was a tendency for a hay by social housing interaction (P=0.08) with GH calves requiring fewer sessions than all other treatments (P<0.03). There was a tendency for a social housing by hay interaction (P=0.07) for test completion time, with IC calves completing sessions faster than the other treatments, suggesting that individually-housed calves may have been more reactive when placed in the T-maze (IC: 27.38, IH: 64.04, GC: 67.9, GH: 74.63 s/session; SE=21.1). These results suggest that both nutritional and social enrichment may increase behavioural flexibility.

**Milk feeding strategy and its influence on development of feeding behavior for dairy calves in automated feeding systems**

*Melissa M. Cantor[1], Amy L. Stanton[2], David K. Combs[3] and Joao H.C. Costa[1]*
[1]*University of Kentucky, Dairy Science Program, Dept. of Animal and Food Sciences, Lexington, KY 40546-025, USA,* [2]*Next Generation Dairy Consulting, Ilderton, ON, N0M 2A0, Canada,* [3]*University of Wisconsin-Madison, Department of Dairy Science, Madison, WI 53706, USA; melissa.cantor@uky.edu*

This study aimed to assess the influence of probiotics and milk feeding strategy on the development of feeding behavior in dairy calves. Ninety-six heifers were enrolled at 7±2 d of age in a 2×2 factorial design study comparing feeding (I) a probiotic, and (II) 2 milk feeding strategies in an automated milk feeding system (AFS). The probiotic contained viable lactic acid fauna and the yeast *Saccharomyces cerevisiae*, and was dispensed by the AFS 2 times per day. We expected probiotics to increase milk intake and growth by influencing positively rumen and lower gut microflora establishment. We expected that an early peak of milk allowance to decrease total milk intake and negatively affect hunger associated behaviors. The early milk feeding strategy (EM) offered a maximum of 11 l on d 1 and peaked at 15 l on d 21. The late milk feeding strategy (LM) offered a maximum of 7 l on d 1, and peaked at 13 l on d 28. Both milk feeding strategies gradually weaned dairy calves after peak milk until d 53 and offered a total of 543 l of pasteurized waste milk. Water and starter were provided *ad libitum*. Milk intake, drinking speed, and the number of rewarded and unrewarded visits were recorded daily by AFS until weaning. Milk intake and rewarded visits were non-normally distributed and analyzed with a Wilcoxon Rank Sum test for the effect of treatment. The effect of treatment on milk feeding behaviors was analyzed using linear mixed models with initial body weight and initial age enrolled on AFS as covariates. A common log transformation was used for the unrewarded visits variable. Probiotic feeding did not affect feeding behavior. All feeding behavior variables interacted with week. Milk intakes were higher during week 1 for EM ($P<0.05$), were not different for week 2 and week 3 ($P>0.10$) and were higher for LM for weeks 4, 5, 6, and 7 ($P<0.05$). Milk intake over the whole experimental period was higher for the LM treatment compared to EM by 67 l ($P<0.001$). Drinking speed was faster for EM over LM at initial daily milk restriction (week 4) ($P<0.001$), and EM was fastest at week 5 over LM (EM=0.9±0.03, LM 0.8±0.03 l/min; $P<0.001$). Number of rewarded visits were not affected by probiotic or milk feeding strategy. Unrewarded visits first differed for week 3 between treatments EM 0.2 visits/d (95% CI: 0.09 to 0.4) and LM 0.3 visits/d (95% CI: 0.3 to 0.5; $P=0.03$). Unrewarded visits were higher for EM over LM for week 4, week 5, week 6 and week 7 ($P<0.001$). Unrewarded visits were highest at week 7 EM 4.5 visits/d (95% CI: 3.9 to 5.3 visits), LM 2.8 visits/d (95% CI: 2.4 to 3.4 visits; $P<0.001$). By week 8, unrewarded visits were not different by milk strategy ($P=0.43$). This research suggests milk feeding strategy influences developmental feeding behavior of calves weeks before weaning, affecting drinking speed, and unrewarded visits. Providing the calf's peak daily milk allotments early in life, followed by earlier milk restriction leads to signs of hunger.

## Consistently efficient: the use of feeding behavior to identify efficient beef cattle

*Ira Parsons, Courtney Daigle and Gordon Carstens*
*Texas A&M University, Department of Animal Science, 2471 TAMU, College Station, TX 77843,*
*USA; ilparsons@tamu.edu*

Increasing the genetic merit of beef cattle for feed efficiency is an effective strategy to improve the economic and environmental sustainability of beef production. Residual feed intake (RFI) is a measure of feed efficiency independent of average daily gain (ADG) and body weight (BW), whereby feed-efficient animals have lower dry matter intake (DMI) than expected (low-RFI). The objective of this study was to examine distinctive feeding behavior (FB) patterns among healthy steers with divergent phenotypes for RFI with the aim to develop the use of FB as a bio-marker of RFI. Three trials were conducted with 508 crossbred steers (BW=309±56 kg) fed a high-grain diet in pens equipped with electronic feeders (GrowSafe® System). Individual DMI, FB traits, and BW, were measured for 70 d, and RFI was calculated as the residual from the regression of DMI on ADG and $BW^{0.75}$. Steers were ranked by RFI and assigned to 1 of 3 RFI classes based on ±0.5 SD from the mean RFI: Low-RFI (n=146), Medium-RFI (n=210), and High-RFI (n=152). For each steer, 17 FB traits were evaluated: frequency and duration of bunk visit (BV) and meal events, head-down (HD) duration, time-to-bunk (TTB; interval from feed delivery to 1$^{st}$ BV event), maximum non-feeding interval, and corresponding day-to-day variation (SD) of these traits. Additionally, 3 ratio traits were considered: BV frequency per meal event, HD duration per meal event and HD duration per BV event. Data analysis was conducted using a mixed-model (SAS 9.4) that included fixed effects of RFI class, year and pen within year. Low-RFI (feed-efficient) steers consumed 16% less ($P<0.01$) DMI, while BW and ADG were not different then High-RFI steers. Compared to High-RFI steers, low-RFI steers had 18% fewer and 24% shorter BV events and 11% fewer meal events that were 13% shorter ($P<0.01$) in length. Compared to High-RFI steers, Low-RFI steers exhibited 10% less ($P<0.01$) day-to-day variation in DMI, as well as 12 to 36% less day-to-day variance ($P<0.01$) in HD duration, BV frequency, BV duration, and meal frequency and duration. Low-RFI steers had 12% longer TTB ($P<0.01$) and were 7% less variable ($P<0.05$) in their TTB compared to High-RFI steers. Partial least squares analysis on these results identified 9 of the 17 FB traits that explained 42% of the variance in RFI. These results illustrate that feed-efficient steers spend less time eating, visit the bunk less frequently, and are more consistent in their FB. Therefore, feeding behavior may be a useful biomarker to identify cattle that are more biologically efficient.

## Multisensorial stimulation promotes affinity between dairy calves and stockperson

*Paula P. Valente[1], Douglas H. Silva Almeida[2], Karen C. Rocha Góis[2] and Mateus Paranhos Da Costa[1]*
*[1]Faculty of Agricultural and Veterinary Sciences, São Paulo State University, Department of Animal Science, Rod. Prof. Paulo Donato Castellane, km 5, 14884-900, Jaboticabal-SP, Brazil, [2]Graduate Program in Animal Science, FCAV-UNESP, Rod. Prof. Paulo Donato Castellane, km 5, 14.884-900, Jaboticabal-SP, Brazil; mpcosta@fcav.unesp.br*

The aim of this study was to evaluate the effect of multisensory stimulation (visual, tactile and auditory) in the behavioral responses of calves. Twenty-four Girolando heifer calves were assessed from birth to 110 days of age. The animals were kept in an individual outdoor housing system from the 2nd to the 80th day of age, where they were stroked twice a day (1 min. each, at milk feeding) during the first 35 days of age. When 47 days old, calves were divided into two groups (G1 and G2), of twelve calves each. Calves from G1 received minimum human contact, and those from G2 were submitted to 5 min. daily of multisensorial stimulation (MSS) from 50 to 80 days of age. MSS was always carried out by the same female stockperson, who approached each calf slowly, making visual contact and talking softly with it (in a low and rhythmic voice), while she caressed its head, neck, dorsal line, flanks, tail insertion, legs, and udder. Calves were weaned on the 70th day of age and moved to paddocks (in groups) ten days later. Some behavioral assessments were performed when the calves were 47, 83, and 93 days of age, by measuring the flight distance (FD, with a non-familiar person), the frequency of vocalization (VOC), and the latency (LAT), frequency (FREQ) and duration (DUR) of the interactions with novel objects. Discrimination tests were carried out when calves were 93 and 110 days of age, by counting the number of calves (NC) that got close to one between two familiar stockpersons, one positive (a woman, responsible for milk feeding and MSS) and one negative (a man, in charge of carrying painful procedures, such as restraining, vaccinations, and medicine injections), who were positioned on the opposite sides of a corridor (4 m wide) where the calves were driven through. Analyses of variance were performed on FD and VOC data, with fixed effects for groups and calf age. Friedman test was used to examine effects of age on LAT, FREQ and DUR; while Mann-Whitney tests were used to analyses the effects of groups (G1 and G2) on these variables. $\chi^2$ test was used to compare the NC between positive and negative stockpersons. Calf age influenced FD ($P=0.03$) and VOC ($P=0.01$), with an increase in FD and a decrease in VOC, over time. No significant differences between groups (G1 and G2) were found for any variable, despite the numerical difference of LAT means (measured at 83 days of age) between the groups, with G2 mean 7.5 times lower than G1. Additionally, a significant increase ($P=0.02$) in the LAT for G2 was found at 83 compared with 93 days of age. The results from the discrimination test indicated that the G2 calves were better able to distinguish between the positive and negative stockpersons, preferring to be close to the positive one ($P=0.03$). Based on these results we concluded that MSS provided an opportunity to create a positive relationship between the calves and stockperson. Financial support FAPESP (Process no. 2015/00606-1).

## The number and outcome of competitive encounters at the feed bunk have an impact on feeding behaviour and growth performance

*Diego Moya[1] and Karen Schwartkopf-Genswein[2]*
[1]*University of Saskatchewan, Large Animal Clinical Sciences, 52 Campus Drive, S7N5B4, Saskatoon, SK, Canada,* [2]*Agriculture and Agri-Food Canada, Lethbridge Research centre, 5403 1 Ave S, T1J4P4, Lethbridge, AB, Canada; diego.moya@usask.ca*

Seventy-four crossbred beef heifers (522.6±42.35 kg BW) were used in a 53-d experiment to assess if the number and outcome of competitive interactions at the feed bunk between individuals had an impact on their feeding behaviour and growth performance. After a 3-wk period of adaptation to the feeding system, heifers were weighed and distributed homogenously into 8 pens (in groups of 8, 9 or 10), each containing 2 feeding bunks equipped with an electronic monitoring system (GrowSafe Systems) for automatic recording of feed intake and feeding behaviour. One out of 4 possible dietary treatments were provided in each pen: a total mixed ration (TMR) (75% barley-grain (BG), 21% corn silage (CS), 4% supplement); or a free-choice diet comprised of either barley-grain and corn silage (BGCS), corn distillers' grain (DG) and barley-grain (BGDG), or corn silage and corn distillers' grain (CSDG). Heifers were also weighed at the end of the study to calculate growth performance and feed to gain ratio. Based on the distribution of the feeding events recorded, competitive interactions were identified as those when two feeding events from two different heifers where registered at the same feeding bunk in less than 5 sec. At each competitive interaction, the outgoing and incoming animals were identified. Two feeding behaviour traits were then calculated individually: activity, as low (LOW), medium (MED), or high (HIG), based on the amount of competitive interactions; and encounters outcome balance (EOB), as displaced (DIS), balanced (BAL), or incoming (INC), based on the ratio incoming:outgoing outcomes from those interactions. Data was analyzed using a mixed-effects model including treatment, activity, EOB and their interactions, and pen was used as the random effect. Heifers with an INC EOB tended to have ($P=0.09$) greater growth performance than those with BAL and DIS EOB. The DIS heifers spent less ($P<0.01$) time with their head down in the feed bunk eating, and had a greater ($P<0.01$) eating rate than INC and BAL. Heifers with HIG activity had a greater ($P<0.01$) frequency of visits to the feed bunk, and tended ($P=0.09$) to spend more time in there compared to LOW heifers. When looking at their behaviour relative to the feed choice offered, INC cattle spent more ($P=0.01$) time with their head down in the feed bunk with BG than DIS, had a greater ($P=0.03$) eating rate of BG than BAL and DIS, and had a lower ($P<0.01$) eating rate of CS than BAL and DIS, while DIS cattle had a lower ($P<0.05$) frequency of visits and consumption of DG than INC and BAL. The HIG heifers had a greater ($P<0.05$) frequency of visits, time at the feed bunk, consumption and percentage of inclusion of DG than MED and LOW. These results suggest that individual animal behaviour described from competitive encounters at the feed bunk changed feeding behaviour towards different feed options with a small influence on growth performance.

**Day-to-day variation in grazing behavior in relation to average daily gain in rangeland beef cattle**

*Matthew M. McInstosh[1], Andrés F. Cibils[1], Rick E. Estell[2], Alfredo L. Gonzalez[2], Shelemia Nyamuryekung'e[1] and Sheri Spiegal[2]*
*[1]New Mexico State University, Dept. of Animal and Range Sciences, Las Cruces, NM 88003, USA, [2]USDA-ARS, USDA-ARS Jornada Experimental Range, Las Cruces, NM 88003, USA; acibils@nmsu.edu*

Increasingly sophisticated movement sensors and data mining algorithms are being used to develop precision grazing tools that monitor livestock behavior in real time. These tools seek to enable livestock producers to adjust management proactively and enhance animal wellbeing and production. A basic challenge associated with these systems is identifying critical behavior parameter values at which to deploy early intervention warnings. This study sought to determine typical ranges of day-to-day variation in grazing behavior of rangeland-raised beef cattle and to explore linear relationships between such variation and average daily weight gain (ADG). Seventeen Raramuri Criollo or Criollo crossbred yearling steers weighing 318±9.3 kg (winter) or 358±8.4 kg (summer) were fitted with Lotek 3300LR GPS collars that recorded animal location at 5 min intervals. Steers grazed a 3,500 ha rangeland pasture for approximately 30 d during either winter (W) or late summer (LS) in 2016 and 2017. GPS data were used to derive 22 commonly-monitored behavior variables. All animal handling protocols were approved by NMSU's IACUC. We calculated daily means and coefficients of variation (CV) of behavior variables for each animal as well as linear correlations between the CV of each behavior and ADG. Day-to-day variation was strongly dependent on the behavior variable considered. Daily time spent resting exhibited the lowest average day-to-day variation in both W and LS (CV=9.9 and 10.8%, respectively) whereas time spent at the drinker (CV=336.6%) and area explored during pre-dawn hours (CV=240.7%) exhibited the highest daily variation in LS and W, respectively. During W, increasing day-to-day variation in distance traveled during the day, or in path sinuosity during pre-dawn hours, or in daily time spent traveling were associated with increasing ADG (r=0.56 to 0.58; $P<0.05$). In LS, steers that exhibited higher CV in 24 h distance traveled or in area explored during pre-dawn hours tended to gain less weight (r=-0.75 to -0.79; $P<0.01$); conversely, steers that showed higher day-to-day variation in path sinuosity or time spent close to the drinker tended to gain more weight (r=0.82 to 0.93; $P<0.01$). Behavior variables that correlated with ADG exhibited intermediate CV values. In this system, either a decrease or increase in day-to-day variation in the expression of a given behavior could signal imminent ADG challenges depending on conditions of the grazing environment (season) and the behavior variable itself.

## Can trees replace the need for wallowing in river buffalo (*Bubalus bubalis*) in the tropics?

*M. Galloso[1,2], V. Rodriguez-Estévez[2], L. Simon[1], M. Soca[1], C.A Alvarez-Diaz[3] and D. Dublin[4]*
*[1]EEPF: Indio Hatuey, Central España Republicana. Matanzas, 44280, Cuba, [2]Córdoba University, Animal Production, Medina Zahara Avenue, 5, 14071, Spain, [3]Universidad Técnica de Machala, UACA, El oro, EC070102, Ecuador, [4]Conservation International, Tokyo, 1800022, Japan; z62gahea@uco.es*

The production systems of buffaloes are receiving increasing attention in the tropics especially in Latin America. The incorporation of trees in pastures (silvopastoral systems) can improve production conditions, animal welfare and increase the availability of food. The objective of the study was to characterize the feeding behavior (FB) of buffaloes in a silvopastoral system (SPS) and a system without trees (SWT), during the Heavy Rain Period (HRP) and the Little Rain period (LRP) where the following activities were recorded during the daylight period, the feeding: grazing, rumination, feeding on tree leaves and water intake; as well as other activities: standing, walking, lying, sheltering and wallowing, to know whether the inclusion of trees in pastures affects thermoregulatory behavior (TB) in these animals. Nine river buffaloes (*Bubalus bubalis*) with an average weight of 90 kg were observed during 12 days between 6:00-18:00 hours and recordings were made at 10-minute intervals. The animals were first observed in the SPS for 12 days and then transferred to the SWT for 12 days during both the HRP and LRP. The Levene statistical test was used for independent samples and ANOVA for the number and frequency of animals in activity per epoch and system studied; the equation of Petit was applied to estimate the time dedicated to each activity. The principal effects to the epoch and systems describe a TB in SPSLRP 0.49 hours and showed significant differences with SWTLRP, 2.83 hours ($P<0.05$); in the HRP, TB showed significant differences 3.70 hours in SWTHRP vs SPSHRP 4.80 hours ($P<0.05$). These results indicate that for the buffaloes in the SPSHRP, 0.31 hours are dedicated to the ingestion of tree leaves and 0.71 hours ($P<0.05$) in the SPSLRP vs 0.10 hours in the SWTLRP ($P<0.05$). There were no differences in the SWT for both seasons. The duration of wallowing was 0.64 hours in SPSHRP vs 4.40 hours in SWTHRP ($P<0.05$), in the LRP there were no statistical differences for this behavior. The FB was different for SPSLRP 6.84 hours and the SWTLRP 7.61 hours, and the SWTHRP dedicated 9.8 hours and SPSHRP 10.47 hours ($P<0.05$). These results describe the behavior and indicate the importance of Silvopastoral systems as technological alternatives friendly to the welfare of animals.

**Evaluation of exercise programs as a management strategy to enhance beef cattle welfare on entry to a feedlot**

*Courtney Daigle, Amanda Mathias, Emily Ridge, Tryon Wickersham, Ron Gill and Jason Sawyer*
*Texas A&M University, Animal Science, 2471 TAMU, 77843, USA; cdaigle@tamu.edu*

Cattle arriving at the feedyard face a myriad of challenges. Testimonial reports describe health and productivity benefits associated with exercising cattle upon arrival at the feedyard, yet little empirical evidence exists to support these claims. To quantify the impact of two exercise programs against a control, two sister studies were conducted simultaneously across 30 days: one in the research (n=210 calves; 12 pens with 4 pens/trt) and one in the commercial setting (n=688 calves; 6 pens with 2 pens/trt). Calves were sorted by sex and placed into pens; treatments were applied to the entire pen 3×/wk for 4 wk and behavioral observations were conducted prior to, during, and after the exercise treatments on days that the cattle were not engaged in their assigned exercise program. Pens were randomly assigned to one of three treatments within sex blocks: (1) programmatic exercise (PRO; cattle moved to drive alley and encouraged to maintain movement for 20 min); (2) free exercise (FREE; cattle moved to drive alley, and allowed free movement without access to the pen for 60 min); or (3) no exercise (CON). Cattle behavior (time budgets, social interactions, exit velocity from the chute), health (vaccine response, health treatments), and productivity (ADG, feed efficiency) were measured across both 30 day-long studies. General Linear Mixed models included a random effect of pen within sex × treatment interaction. Results were similar for both studies. No measurable production ($P>0.06$) or health ($P>0.38$) benefits were identified for either PRO or FREE calves compared to CON calves. Calf behavior was not impacted by treatment ($P>0.55$). Risks associated with exercising calves included dust generation, injury to cattle, handler, and working equids, and infrastructure damage and repair. No production or health benefits were observed; managers should evaluate risks to cattle prior to implementing an exercise program at their facility.

### Effect of acclimation and low-stress handling on beef cattle activity

*Rebecca Parsons[1], Grant Dewell[1], Reneé Dewell[2], Johann Coetzee[3], Tom Noffsinger[4], Anna Johnson[5] and Suzanne Millman[1,6]*
[1]*Iowa State University, Veterinary Diagnostic & Production Animal Medicine, Ames, IA 50011, USA,* [2]*Iowa State University, Center for Food Security & Public Health, Ames, IA 50011, USA,* [3]*Kansas State University, Department of Anatomy & Physiology, Manhattan, KS 66506, USA,* [4]*Production Animal Consultants, Oakley, KS, 67748, USA,* [5]*Iowa State University, Department of Animal Science, Ames, IA 50011, USA,* [6]*Iowa State University, Biomedical Sciences, Ames, IA 50011, USA; bparsons@iastate.edu*

Weaning, transport and arrival to feedlots is stressful for calves and increases the risk of morbidity and mortality. Acclimating cattle to the feedlot and low-stress handling (utilizing flight zones and body pressure) may reduce stress. The effect of acclimation on cattle activity (Motion Index [overall activity measurement; MI], percentage time lying and number of steps taken) was assessed using accelerometers. We hypothesized that acclimated cattle would have lower MI, spend more time lying, and take fewer steps than controls. Upon arrival to the feedlot, cattle were assigned to one of two treatments by pen, control or acclimated (about 50 cattle per pen; 903 acclimated and 905 control cattle). During acclimation (about 45-60 min total), the cattle were encouraged to form a cohesive herd in the home pen, moved as a herd around the pen and then moved through the vaccination shed and chute. Upon return to the home pen, cattle were encouraged to move to the waterer and feed bunk. Acclimation occurred once daily, beginning the day after arrival and 2 more times following initial vaccinations (Day 1), but within the first 5 days after Day 1. During the acclimation period, control cattle were not handled by the researchers. On Day 1, all cattle were vaccinated. IceTag accelerometers were placed on the left hind leg of random sentinel animals (acclimation, n=28; control, n=38). All cattle were revaccinated at approximately Day 10 and IceTags removed. PROC GLIMMIX of SAS was used to assess differences in activity on Days 1-8 between acclimated and control cattle. The MI was averaged for each day, while step number was summed for each day. The model included the interaction of Day and Treatment. On Day 1, acclimated cattle had lower MI than control cattle (4.96±0.5 and 6.86±0.6, respectively; $P=0.03$). Within the acclimation treatment, Day 1 MI was higher than Days 3 and 4 (4.96±0.5, 3.51±0.5 and 3.27±0.4, respectively; $P<0.05$). Within the control treatment, Day 1 MI (6.86±0.6) was higher than all other days (range: 3.45±0.5 to 3.97±0.5; $P<0.01$). Lying behavior was not different between control and treatment cattle or across days ($P=0.98$). No differences between treatments in step number were observed ($P>0.05$). There was no difference in step number by control cattle across days (range: 1,600.07±173.4 to 2,035.96±173.4 steps; $P>0.05$); however, number of steps by acclimated cattle on Day 1 (1,353.46±173.4) was lower than all days (range: 1,979.04±173.4 to 2,287.25±173.4; $P≤0.01$), except Day 4 (1,825±173.4; $P>0.05$). Lower MI on Day 1 in acclimated cattle may indicate they were calmer and more able to recover from vaccination stress than control cattle. The increase in the step number in the acclimated cattle may be attributed to the acclimation treatment that took place on the days following Day 1.

**Effects of running water on drinking behaviour by *Bos taurus* cattle being rested during long-distance transportation**

*Derek Haley[1,2], Ray Stortz[1,3], Michael Ross[1,3] and Tina Widowski[1,3]*
*[1]University of Guelph, Campbell Centre for the Study of Animal Welfare, 50 Stone Rd E, Guelph, ON, N1G 2W1, Canada, [2]University of Guelph, Population Medicine, Ontario Veterinary College, 50 Stone Rd E, Guelph, ON, N1G 2W1, Canada, [3]University of Guelph, Animal Biosciences, Ontario Agricultural College, 50 Stone Rd E, Guelph, ON, N1G 2W1, Canada; dhaley@uoguelph.ca*

Current transport regulations in Canada require that cattle on a road journey expected to exceed 48 h must be unloaded for feed, water, and rest, for a minimum of 5 h. Previous work has reported that the first behavioural priority for cattle unloaded for rest in transit, was eating. In those studies, after unloading, cattle were moved into pens containing a large round hay-bale feeder right in the middle of their pen. Thus, they may have eaten first as it was the first option presented, rather than due to a higher motivation to eat (vs drink), *per se*. The objective of this study was to test whether having the visual and auditory stimulation from water running into the water trough would impact the drinking behaviour of cattle unloaded for feed, water, and rest during long-distance transportation. Twenty-nine truckloads of cattle were observed at a commercial rest station near Thunder Bay, Ontario, and for each load cattle were divided into a control and treatment groups. As is conventional practice, none of the cattle had access to drinking water while on the transport truck. Control groups were presented with aerated water in a cement bunker trough as per standard rest station conditions, while treatment groups were presented with water running into the trough by use of a submersible pump which discharged water at a height of 0.56 m above the trough. The number of animals drinking, eating, lying and performing 'other' behaviour was recorded every 5 min for the first 5 h after cattle were unloaded. The running water treatment had no significant effect on drinking behaviour ($P=0.9395$), nor did it affect eating ($P=0.4925$), lying ($P=0.4207$), or other behavior ($P=0.3792$). A significant treatment by hour interaction was found for all behaviour patterns ($P\leq0.0001$), as well as a significant treatment by weight class interaction for drinking, eating and lying behavior ($P=0.0001$). Our results showed that cattle were not drawn to drinking by the presence of running water at the trough more than aerated water. This study highlights a need for further understanding of drinking behaviour by cattle; specifically how cattle locate drinking sources. Additional information could help to improve the practice of water provision, potentially improving animal health, welfare, and production.

**Evaluation of the effect of a recombinant vaccine for immunocastration in bulls on behavior and animal welfare**

*Paula R. Huenchullán[1] and Leonardo Saenz[2]*
*[1]Universidad de Chile, Doctorado en Cs. Silvoagropecuarias y Veterinarias, Santa Rosa 11315, 8820808, Santiago, Chile, [2]Leonardo Saenz, Fac. Cs. Veterinarias, Santa Rosa 11735, 8820808, Santiago, Chile; prhuenchullan@gmail.com*

Immunocastration or vaccination against GnRH is a tool that has been studied as an alternative to surgical castration, with the aim of performing reproductive and behavioral control under animal welfare standards. The objective of this study was to evaluate the effect of two adjuvants based on proteoliposomes (P) and chitosan (Q) with different immunostimulating properties, in a bovine immunocastration vaccine that allows the generation of a vaccination strategy with as few doses as possible. Two clinical trials were conducted in different commercial systems (S1 and S2). Each trial consisted of 120 entire males that were divided into 3 groups each composed of 40 animals. The systems were constituted in the following way: _ S1: control group, surgically castrated (CQ); group 2, Immunocastrated with proteoliposomes as adjuvant (P); group 3, Immunocastrated with chitosan as adjuvant (Q). _ S2: control group, entire males (CE); group 2, Immunocastrated with proteoliposomes as adjuvant (P); group 3, Immunocastrated with chitosan as adjuvant (Q). The vaccine was evaluated through behavior, blood and production variables. The blood samples were taken on days -30, +37, +75, +122, 164 and +194 (S1) and -5, +26, +85, +184 and +323 (S2) to measure testosterone concentrations and antibodies against GnRH, glucose, CK and the neutrophil ratio: lymphocytes for all groups. In the direct observation frequencies of the agonist, sexual, social, exit speed and temperament score were recorded (S1 and S2). The production and meat quality indicators were evaluated in S1 and only behaviour and some production parameters in S2. The sensory analysis was carried out through a panel of experts from the Faculty of Agronomy of the University of Chile. Significance and statistical differences between the groups of each system were determined by ANOVA and Scheffe tests (a posteriori) in the case of the variables that were normally distributed. For the variables that did not have a normal distribution, Kruskal-Wallis plus Bon Ferroni was applied. Values of $P \leq 0.05$ were considered statistically significant. In S1, IC animals with chitosan as adjuvant showed a significant difference during the entire study period in Ac anti GnRX G/Q levels with respect to the other groups. In addition, testosterone levels between the CQ and Q group were not significantly different during the study. The averages of the measurements of sexual, agonist and affiliative behavior did not show differences between the groups. However, group P came out significantly slower than the control group at the end of the trial period. With respect to the production indicators, the immunocastrated groups (P and Q) obtained greater slaughter weight, hot carcass weight and yield ($P \leq 0.05$).

## An investigation of behaviours displayed by pre-weaned dairy calves in different group sizes on commercial Irish dairy farms

*John Barry[1,2], Eddie A.M. Bokkers[1], Imke J.M. De Boer[1] and Emer Kennedy[2]*
*[1]Wageningen University & Research, Animal Production Systems Group, P.O. Box 338, 6700 AH, Wageningen, the Netherlands, [2]Teagasc Moorepark, Animal & Grassland Research and Innovation Centre, Fermoy, Co. Cork, Ireland; john.barry@teagasc.ie*

Calves are social animals and research has shown that both welfare and performance can be improved when calves are group-housed. In Ireland, spring calving systems predominate, targeting >90% calving rate in a 6-week period. This creates challenges, particularly with regard to labour requirements when managing calves. Grouping of calves, from 3 days of age, is commonly practiced to reduce the required labour, however, due to the volume of calves born in a short period, large groups with >20 calves are not uncommon. Cohabitating of large numbers of calves, of similar ages, albeit within the legally required legal space allowance ($\geq 1.5$ m$^2$/calf), could affect welfare as the risk of disease transmission is increased, but also the expression of behaviours, such as resting and playing, could be impeded in what is potentially a stressful environment. This study investigated the effect of group size, when space allowance was $\geq 1.5$ m$^2$/calf, on calf behaviour on commercial Irish dairy farms operating a seasonal calving system. Commercial dairy farms (n=47), located in the Munster region of Ireland were visited in spring (February – April) 2017. Video cameras were setup to record, without disturbance, two separate groups for 60 minutes (two farms per day were observed between 10:00 and 16:00 h). All calves observed were under six weeks of age. Recordings were later analysed by scan sampling at 5-minute intervals ($\pm 30$ seconds) using a detailed ethogram. Effect of group size on resting, playing, standing, walking and grooming behaviour was investigated using a mixed model. Group nested within farm was included as a repeated measure. Group size was categorised from 1 to 3, where 1 = <10 calves; 2 = 10-15 calves, and 3 = >15 calves per category. Category 1 contained 36 groups, and category 2 and 3 contained 34 and 22 groups respectively. Mean group size was 11.4 (SD 5.98) calves, and ranged from 3 to 31 calves. Group size affected play behaviour frequency ($P<0.05$), where category 1 had higher mean playing frequencies (4.49%) than category 2 (1.83%: $P<0.01$) and category 3 (1.88%: $P<0.01$). Group size had no effect on the frequency of resting (72.4%; $P=0.67$), standing (17.3%; $P=0.57$), grooming (1.9%; $P=0.11$), walking (1.9%; $P=0.71$), and eating (4.89%; $P=0.41$) behaviour, observed within the 60-minute period. Behaviour observations, from scan sampling 60-minute video recordings, show higher levels of play behaviour by calves in group sizes of <10. Since this behaviour is generally associated with good welfare, limiting group sizes to $\leq 10$ calves might promote good welfare, in seasonal calving systems.

**The effects of housing dams and calves together on behaviour, growth, and production**

*Amanda R. Lee[1], Ashley D. Campaeux[1], Liesel G. Schneider[1], Melissa C. Cantor[2], Joao H.C. Costa[2] and Peter D. Krawczel[1]*
[1]*The University of Tennessee, Animal Science, 2506 River Drive, Knoxville, TN 37996, USA,* [2]*University of Kentucky, Animal and Food Sciences, 900 W.P. Garrigus Building, Lexington, KY 40546, USA; krawczel@utk.edu*

Early separation of calves from dams is a contentious issue in the dairy industry. Because cows and calves naturally remain together, early separations add to consumers' negative stereotypes of the dairy industry. The objective of this study was to evaluate differences in calf and dam behaviors among calves and dams group housed nightly (20:00 to 6:00) and calves and dams housed together, nightly. Cow-calf pairs (n=20) were separated within 6 h of calving and randomly assigned to either: (I) group calf housing on pasture nightly and dams housed on sand bedded freestalls 24 hours per day, or (II) dams and calves housed together on pasture nightly, from days 5 to 19 postpartum. Five-day separation was used to ensure successful passive transfer (serum total protein ≥5.5 between days 1 and 3) and no clinical symptoms of mastitis prior to cow-calf enrollment. Calves were fed 4 l per day of 26% crude protein, 20% fat calf milk replacer mixed with water until day 4 and 6 l per day from days 5 to 19. Accelerometers were attached to dams and calves between days 3 and 5 postpartum. Somatic cell score (SCS) and calf body weight were collected weekly. Calf approachability was scored on day 10±3 with 1 being unable to approach within 2 m of calf to 10 being calf follows researcher post-testing. After day 19, calves were permanently separated from their dams and group housed 24 hours per day on pasture until weaning, and dams were housed exclusively in sand-bedded freestalls. An ANOVA was conducted using the MIXED procedure (SAS 9.4) to evaluate the effects of treatment and calf sex on SCS and dam and calf lying time and bouts. T-tests were conducted to evaluate the effect of treatment on average daily gain (ADG) and approachability. Somatic cell score did not differ between dams housed with and without their calves, respectively (2.84±0.53, 2.01±0.47; P=0.25). Dams housed with calves nightly, averaged 319 more steps per day than dams housed indoors (P=0.01). Calf sex was not a significant predictor of SCS, lying bouts, or steps per day of calves and dams (P≥0.06). Calf ADG and approachability did not vary between treatment groups or sex (P≥0.20). Housing dams and calves did not affect calf ADG, dam and calf activity, or SCS, suggesting housing dams with calves after day 5 may provide one option to address consumers' concerns without negative consequences to the dam or calf.

**Consistency of personality traits from calf to adulthood in dairy cattle**

*Heather W. Neave, Joao H.C. Costa, Daniel M. Weary and Marina A.G. Von Keyserlingk*
*University of British Columbia, Animal Welfare Program, Faculty of Land and Food Systems,*
*2357 Main Mall, V6T 1Z4, Canada; hwneave@gmail.com*

Little is known about the consistency of personality traits from calf to adulthood. The aim of this study was to determine the long-term consistency of individual personality traits from pre-weaning through to first lactation. Calves were enrolled in the study in two cohorts and subjected individually to 3 tests designed to measure behavioural reactivity to novelty at 4 different ages: pre-weaning (age 26 d, Cohort 1: n=32), post-weaning (age 76 d for Cohort 1: n=32; age 100 d for Cohort 2: n=33), heifer (age 12 mo, Cohort 2: n=31), and cow (age 30 mo and pregnant in first lactation; Cohort 1: n=14, Cohort 2: n=20). The personality tests conducted at each age included: novel environment (30 min in an unfamiliar arena bedded with sawdust), human approach (10 min with an unfamiliar stationary human) and novel object test (15 min with an unfamiliar object). Behaviours of each individual in each test were analysed for consistency in the same test across age using Spearman's rank-order correlation. All five behaviours recorded in the human approach test (latency to touch, time spent looking at, touching, or playing, and number of object play bouts) were consistent between pre-weaning and post-weaning periods, and between heifer and cow periods (r>0.30; $P<0.05$). Similar consistency in behaviours in the novel object test were found between the pre- and post-weaning periods (time spent looking at or playing, latency to touch; r>0.35; $P<0.05$) and between heifer and cow periods (touching and looking; r>0.44; $P<0.05$). All three behaviours recorded in the novel environment test (time spent exploring, active, or inactive) were consistent between heifer and cow periods (r>0.54; $P<0.05$), but not between the pre- and post-weaning periods. There was little consistency in any of the behaviours between the post-weaning and heifer periods. These results suggest that personality traits in dairy cattle are generally stable within the early and later rearing periods, but are less consistent across these periods perhaps due to developmental changes.

## Evaluation of single vs paired calf housing on the standing behavior of Holstein dairy calves

*Clay Kesterson[1], Liesel Schneider[1], Marc Caldwell[2], Gina Pighetti[1] and Peter Krawczel[1]*
[1]*The University of Tennessee, Knoxville, 2506 River Drive, 37996 Knoxville, TN, USA,* [2]*The University of Tennessee College of Veterinary Medicine, 2407 River Drive, 37996 Knoxville, TN, USA; ckesters@vols.utk.edu*

Pair housing dairy calves during the pre-weaning phase can improve feed intake following weaning and cognitive function prior to weaning. However, it is not established what behavioral or performance differences resulting from social housing may lead to later success. Our objective was to determine the effect of pair vs individually housed calves on standing behavior and growth. At 5±1 d relative to birth, calves with successful passive transfer of immunoglobulins from colostrum (STP reading >5.5 g/dl) were blocked by sex and birth date and assigned to either pair (n=14) or individual (n=14) housing. Calf pairing was implemented by combining two individual pens. One paired calf served as the focal calf and the other imposed treatment. All data were collected from the focal calf in pair housing. Control calves remained individually housed. Milk replacer (protein 26%: fat 20%; 3 l) was fed 2×/d and grain and water were provided ad libitum. Pens were deep-bedded with straw and located underneath an open-walled structure. Accelerometers were affixed on the calf's rear leg on d 8±1, relative to birth, to monitor standing time. The accelerometers were shifted to the opposite leg every 7 d to prevent lesions. Body weight was recorded weekly, at weaning, and 1 week post-weaning. A MIXED procedure (SAS 9.4) was used to run multiple ANOVAs to evaluate the effects of housing treatments on standing time, bouts, duration, and bout duration, and growth. Random effects of calf, birthdate, sex, and the repeated measures of calf and sampling within days were included in all models. Paired calves had greater standing time in relation to individually housed calves ($P<0.001$; paired 6.0±0.3 vs single 4.7±0.3 h/d). However, standing bouts were similar between housing treatments ($P=0.09$; paired 19.8±1.1 vs single 17.4±1.1 bouts/d). Bout duration tended to be longer for the paired calves ($P=0.07$; paired 20.1±1.6 vs single 17.0±1.5 min/d). Mean body weight gain ($P=0.35$) and calf growth rates from weaning to one-week-post-weaning were not different between treatments ($P=0.67$). Additionally, no effect of treatment on body weight gain occurred one week post-weaning. Although weight gain and standing bouts did not differ between single and paired calves, paired calves increased their standing time, and tended to have more standing bouts. The increased standing time was likely driven by longer bouts of standing. This suggests that the increased standing time was driven by more time engaging in playful, explorative, and/or feeding behaviors relative to individual calves.

## Cross-sectional study of housing and risk factors for leg dirtiness of smallholder dairy calves in Kenya

*Emily Kathambi[1], John Vanleeuwen[1], George Gitau[2] and Shawn McKenna[1]*
*[1]University of Prince Edward Island, Health Management, 550 University Ave, Charlottetown, C1A 4P3 Prince Edward Island, Canada, [2]University of Nairobi, Faculty of Veterinary Medicine, Clinical Studies, Kabete, Nairobi, Kenya; slmckenna@upei.ca*

The welfare and health of calves are affected by housing management practices. This study was aimed at describing the calf housing and determining the individual and pen level factors that affected calf leg dirtiness of smallholder dairy farms in Meru County, Kenya. A cross-sectional study was carried out on 52 calves that were one year old and younger in 38 dairy farms (mean ± SD: herd size=1.71±0.7 milking cows; milk production=6.7±3.1 l/day) in Meru, Kenya in 2017, to describe their housing and determine the factors associated with leg dirtiness. Calf biodata, health status and leg dirtiness were assessed through physical exams. Pen floor surface area was measured, and pen characteristics were scored, such as floor hardness and dirtiness, through standard knee impact and knee wetness tests, respectively. A questionnaire was administered to the farmers to gather information regarding calf housing management practices on the farm. Univariable and multivariable logistic models were used to determine factors associated ($P<0.05$) with calves with dirty legs (dirtiness score >2.5), controlling for confounding of other variables in the model. With less than 2 calves per herd, farm was not included as a covariate. Heifers formed 52% of the study population, and ages ranged from 1 week to 12 months, with an average of 5.2±3.1 months. Indigenous breeds formed 15% of the study group, while the rest of the calves were primarily crossbreds with small amounts of indigenous blood. Two calves had pneumonia, while the rest appeared healthy. The calves had a mean body weight of 85.2±32.8 kg, and based on their age, current weight, and estimated birth weight, we estimated a mean average daily weight gain (ADG) of 0.68±0.698 kg/day. A total of 71% of calves had a good body condition score greater than or equal to 2.5, and each calf had a space allowance of 2.52±1.56 m². Approximately 75% of the calves (39/52) were kept in pens and the rest were reared outdoors. For the 39 calves kept in pens, 23 and 33% of them had a failed knee impact test and failed knee wetness test, respectively. We observed 62% of pens having bedding, and 26% (10/39) of pen floors being wood or concrete, with 40% (4/10) of these pens having bedding. In univariable logistic regression analyses of the 52 calves, housing calves indoors had an increased odds of legs being dirty by 6.3 times ($P=0.045$), compared to outdoor-reared calves. In the final multivariable logistic regression model of 39 calves in pens, concrete or wood floors (OR=7.9, $P=0.047$), poor body condition score below 2.5 (OR=17.1, $P=0.020$) and use of bedding (OR=12.5, $P=0.046$) were risk factors associated with dirty calf legs, compared to dirt floors, good body condition, and no bedding, respectively. The calves with dirty legs were likely experiencing impaired calf comfort and animal welfare. Smallholder dairy farmers in Kenya should be trained on calf housing and bedding management to improve welfare.

## Individual maternity pens offer protection from disturbances by alien cows

*Margit Bak Jensen[1] and Maria Vilain Rørvang[1,2]*
[1]*Aarhus University, Department of Animal Science, Blichers Allé 20, 8830 Tjele, Denmark,* [2]*Swedish University of Agricultural Sciences, Biosystems and Technology, Sundvägen 6, 23053 Alnarp, Sweden; margitbak.jensen@anis.au.dk*

Use of individual maternity pens is recommended to minimise disturbance during calving. Moving dairy cows to an individual maternity pen prior to calving is a challenge because the time of calving is difficult to predict. However, housing cows in a group area with access to individual maternity pens, which the cows can enter when motivated to isolate, may solve this problem, especially if the pens are fitted with a gate preventing alien cows from following. The aim of this study was to assess the effects of design of a maternity pen on maternal behaviour and disturbance from alien cows during the first hour after calving. Danish Holstein cows (n=78) were housed in 13 groups of six cows in a group area with access to six individual maternity pens. Groups were alternately assigned to a group area where individual pens had a mechanical gate allowing only one cow access at a time (Gate; 7 groups), or a group area where the gate to the individual pens were permanently tied open (No gate). Data for the first hour after birth of the calf was collected from video recordings of 60 multiparous cows with an easy and unassisted calving (32 cows with access to individual pens with a gate and 28 cows with access to an individual pen with no gate). In groups where individual maternity pens had a gate, fewer cows calved in the individual pens compared to in groups with no gate (13/32 vs 19/28; $\chi^2$=4.45; P<0.05). The effect of treatment (Gate, No gate) and calving place (Pen, Group area) on continuous behavioural variables was analysed by a mixed model including the fixed effects of treatment, calving place, treatment × calving place and lactation number, while group was included as a random effect. Cows spent 39 min of the first hour licking their calves, irrespective of whether they calved in an individual pen, or if this pen had a gate. When calving in an individual pen with a gate, the cows tended to interact less with an alien cow (0.01, 4.33, 3.76 and 7.73 times/h for Gate/Pen, Gate/Group area, No gate/Pen and No gate/ Group area; $F_{1,43}$=4.04; P=0.05). Furthermore, when a cow calved in a gated individual pen, a lower disturbance of the calf by an alien cows was observed (0.50, 28, 18 and 33 min/h for Gate/Pen, Gate/Group area, No gate/Pen and No gate/Group area; $F_{1,43}$=6.8; P<0.05). Eleven calves suckled an alien cow within an hour of birth. More of these calves were born in the group area (9/28 calves born in group area vs 2/32 calves born in individual pen; $\chi^2$=6.67; P<0.01), but there was no effect of treatment on this. The results suggest that calving in an individual maternity pen with a gate that prevents other cows from entering, protects dam and calf from disturbance from alien cows. The results also illustrate that being born in a group maternity pen increases the risk of mis-mothering.

**The effect of long-term exposure to concrete or rubber flooring on lying behavior in cattle**

Emma C. Bratton[1], Susan D. Eicher[2], Jeremy N. Marchant-Forde[2], Michael M. Schutz[3] and Katy L. Proudfoot[1]

[1]The Ohio State University, Veterinary Preventive Medicine, 1920 Coffey Rd, Columbus, OH 43210, USA, [2]USDA-ARS, Livestock Behavior Research Unit, 125 S Russell St, West Lafayette, IN 47907, USA, [3]Purdue University, Animal Sciences, 125 S Russell St, West Lafayette, IN 47907, USA; bratton.64@osu.edu

Many dairy cows are housed on concrete flooring due to its ease of manure removal and low cost, despite softer alternative flooring surfaces such as rubber. Standing on harder flooring such as concrete has been associated with lameness, and may also impact a cow's ability to stand up and lie down comfortably. The aim was to determine the effect of concrete or rubber flooring on lying behavior in lactating dairy cows over their first two lactations. Forty pregnant heifers (n=20/flooring type) were enrolled in the study before their first lactation and were randomly assigned to one of two treatment pens which they entered after they gave birth: one pen included grooved concrete and the other included rubber flooring. Both pens included free-stalls with air mattresses on concrete bases, and were dynamic as cows entered and left the pen based on their calving dates. At calving, cows were fitted with 3D accelerometers that automatically recorded daily lying time, number of lying bouts, and lying bout duration across their first two lactations. Due to disease and missing data, accelerometer data from 15 cows on concrete and 14 cows on rubber were included in the final analysis. Eight cows per treatment were video recorded on approximately day 90 of their second lactation to measure the time it took them to transition from standing to lying, and vice versa, averaged across at least 3 bouts/36 h period. All research conformed with the ISAE Ethical Guidelines. To determine the effect of treatment on lying behaviors, a mixed model was used. The model included treatment as a fixed effect, lactation (1 and 2) as a repeated measure, a treatment by lactation interaction, and cow as a random effect. T-tests were used to determine differences between treatments in the time to transition from lying to standing (and vice versa). There was no effect of treatment on mean daily lying time across both lactations (rubber vs concrete: 9.9 vs 9.7±0.3 h/d; $P=0.59$). Cows on rubber flooring had fewer (10 vs 12±0.5 bouts/d; $P=0.025$) and longer lying bouts (63.7 vs 52.5±2.3 min/bout; $P=0.002$) per day compared to cows on concrete, regardless of lactation. There was a trend approaching significance for cows on concrete to take longer to transition from standing to lying (5 vs 4±0.3 sec/bout; $P=0.12$), but there was no effect of flooring on the time it took cows to transition from lying to standing (4 vs 4±0.3 sec/bout; $P=0.73$). Results suggest that cows alter their lying behavior depending on the flooring surface, even when the lying surface remained the same. The longer and less frequent lying bouts for cows housed on rubber may indicate those cows were more comfortable than those on concrete, although overall lying time was low for both groups.

**Stepping behavior and muscle activity of dairy cattle standing on concrete or rubber flooring for 1 or 3 hours**

*Karin E. Schütz[1], Eranda Rajapaksha[2], Erin M. Mintline[3], Neil R. Cox[1] and Cassandra B. Tucker[3]*
*[1]AgResearch Ltd., Ruakura Research Center, Hamilton, New Zealand, [2]University of Peradeniya, Department of Veterinary Clinical Science, Faculty of Veterinary Medicine and Animal Science, 20400, Sri Lanka, [3]University of California, Center for Animal Welfare, Department of Animal Science, 1 Shields Ave, Davis, CA 95616, USA; cbtucker@ucdavis.edu*

The type of flooring in dairy cattle systems influences their health and welfare. While concrete is common, the use of more compressible surfaces, such as rubber, is on the rise. Cows prefer to stand and walk on rubber surfaces compared to concrete, however, it is largely unknown how standing for longer periods of time influence muscle activity and fatigue. Therefore, measures of behavior and muscle activity were used to investigate potential benefits of providing a rubber flooring surface to dairy cattle. Sixteen lactating Holstein cows were forced to stand on either a concrete or rubber surface for 1 or 3 h, in a 2 by 2, cross-over design. Surface electromyograms (SEMG) and skin surface temperature were used to evaluate muscle activity, fatigue, and movement of muscle activity between hind legs. Muscle activity of 2 muscle types, the biceps femoris and middle gluteal, was assessed during both static contractions when cows transferred weight to each hind leg, before and after 1 and 3 h of standing, and dynamic contractions associated with both steps and shifts in weight without steps. In addition, stepping rate, time between each step, feeding behavior, skin surface temperature, and latency to lie down afterwards were evaluated and treatments were compared with mixed models. Standing duration influenced both the behavior and muscle activity of cows. Stepping rate increased with standing time for cows on both flooring types ($P<0.05$). Muscle activity parameters of the biceps femoris muscle were higher after 3 h of standing for cows standing on both flooring types (2.3 and 3.6% increase in median amplitude and median power frequency, respectively) compared to the change after 1 h, both compared to baseline values before the standing treatment ($P<0.03$). The surface type influenced the behavior and muscle activity of the cows during the first hour of exposure only; cows standing on rubber had higher stepping rate, shorter interval between steps and higher number of SEMG shifts (muscle activity shifts with or without visible steps) compared with cows on concrete ($P<0.05$). There was no difference in skin surface temperature, feeding behavior or latency to lie down between the treatments. The results show that standing on a rubber surface causes a different initial behavioral response compared to standing on concrete, however, possible reasons for these changes are unclear. The results regarding muscle activity and muscle fatigue are inconclusive and warrant further investigation.

## Cross-sectional study of cow comfort and risk factors of lying time and cleanliness of smallholder dairy cows in Kenya

*Emily Kathambi[1], John Vanleeuwen[1] and George Gitau[2]*
*[1]University of Prince Edward Island, Health Management, 550 University Ave, Charlottetown, C1A 4P3 Prince Edward Island, Canada, [2]University of Nairobi, Faculty of Veterinary Medicine, Clinical Studies, Kabete, Nairobi, Kenya; jvanleeuwen@upei.ca*

Lying behavior and cow cleanliness are indicators of cow comfort and can be influenced by stall parameters and management practices. This was a cross-sectional study aimed at determining the management practices affecting lying behavior in smallholder dairy cows in Meru, Kenya in 2017. A total of 106 milking cows, predominantly exotic in breed (94%), from 73 farms were assessed for daily lying time, number of lying bouts per day, and duration of each bout, in addition to daily milk yield and presence of mastitis through the California Mastitis Test (CMT). Data loggers (HOBO Pendant G Acceleration Data Logger) attached on the inside of the left hind leg were used to record the lying and standing information of the cows for three days, and the data were exported to Excel® and imported for further analysis into Stata 14®. Leg, udder, and stall hygiene were assessed, along with knee wetness and impact tests, and housing design problems potentially contributing to cow comfort issues. Information on management practices and milk production were acquired using a questionnaire that was administered face-to-face to the farmers in their native Kimeru language. Univariable and multivariable linear and logistic models were used to analyse factors associated ($P<0.05$) with lying behaviour and cow cleanliness, respectively, controlling for confounding of other variables in the models. The mean daily milk yield per cow was $6.61\pm3.32$ litres, and 44 cows (42%) and had a CMT score of $\geq1$. The mean daily lying time, number of bouts and duration of each bout were: $10.9\pm2.2$ hours; $21.0\pm19.1$ bouts per day; and $0.85\pm0.47$ hours, respectively. Knee impact and knee wetness tests failed in 13 and 11% of the stalls, respectively. A total of 65% of stalls were categorised as dirty (>2.5), while 21 and 50% of the cows had udder and leg hygiene scores >2.5, respectively. From multivariable models, at the cow-level, stall hygiene scores >2.5 ($\beta=-1.12$, $P=0.005$), and poorly positioned neck rails ($\beta=-1.505$, $P=0.011$) were associated with decreased lying time of cows, and two farm-level variables were also in the final model: delayed removal of manure ($\beta=-1.534$, $P=0.001$) and delayed addition of new bedding ($\beta=-1.216$, $P=0.014$). Delayed cleaning of the alley (OR=6.1, $P=0.033$), lack of bedding on the stall floor surface (OR=4.97, $P=0.008$) and standing idly and/or backwards in the stall (OR=10.83, $P=0.010$) were herd-level risk factors for stall dirtiness. Stalls categorized as dirty (OR=3.38, $P=0.038$) and adding new bedding less than once a week (OR=3.31, $P=0.041$) were cow- and herd-level risk factors for dirtiness of the udder, respectively, while the stall being dirty was a cow-level risk factor for leg hygiene scores >2.5. Stall design and management practices have a significant impact on cleanliness of cows and their lying behaviour.

### Factors affecting lying behavior of grazing dairy cows in an organic system

*V.L. Couture[1], P.D. Krawczel[1], S.R. Smith[2], L.G. Schneider[1], A.G. Ríus[1] and G.M. Pighetti[1]*
*[1]University of Tennessee, Animal Science, 2506 River Drive, Knoxville, TN 37996, USA, [2]University of Kentucky, Plant and Soil Science, 1405 Veterans Drive, Lexington, KY 40546, USA; vcouture@vols.utk.edu*

Lying behavior is often used in the assessment of dairy cow welfare in confinement systems. However, USDA organic dairy cows must receive >30% DMI from pasture, although there is flexibility in grazing 30 – 100% DMI and utilizing different types of housing. Because of additional time spent foraging, grazing cows may have unique time budgets. Therefore, research examining their lying behavior is needed to offer recommendations for organic and grazing systems. The objective was to quantify lying time and identify variations in the lying behavior of grazing dairy cows under organic management. Lactating dairy cows (n=230) from certified organic, grazing dairy farms (n=5) were enrolled. Management systems were categorized based on housing system and feeding management. Low input (LI) systems (n=3 farms; 171 cows; 188.6±92.8 DIM) utilized a loose housing system and relied on pasture for >50% DMI, while high input (HI) systems (n=2 farms; 59 cows; 197.9±90.5 DIM) used tie-stalls and 30-50% DMI was received from pasture. Accelerometers were affixed to the cows' rear leg for three 28-d periods during the spring (P1), summer (P2), and fall (P3) at the LI farms and during P1 and P3 at the HI farms. Data were analyzed using the MEANS and MIXED procedures in SAS (v9.4). A linear mixed model was developed using backward manual elimination to test the effects of milk yield, parity, and DIM on lying duration in LI and HI systems. Cows on the HI farms laid 11.16±0.06 h/d, while cows on LI farms laid for 8.49±0.03 h/d. Lying time on LI farms changed from P1 to P3 (7.41±0.07 vs 9.21±0.05 h/d), but remained relatively consistent on HI farms (P1=11.13±0.08 h/d; P3=11.20±0.09 h/d). The majority (72.6±3.27%) of lying behavior on LI systems occurred at night from 19:00 – 7:00. On HI farms 46.5±2.1% of lying time took place at night. Furthermore, milk yield, DIM, and parity had an effect on lying time on LI farms ($P \leq 0.001$). Increased milk yield was associated with decreased lying time and as DIM increased, so did lying time ($P \leq 0.001$). Primiparous cows laid less (7.62 h/d) than second (8.85 h/d; $P=0.001$) or third (9.3 h/d; $P \leq 0.001$) parity cows, but were not different from cows in their fourth or greater parity (8.21 h/d; $P=0.44$). However, cows in fourth or greater parity laid less than cows in third parity ($P=0.02$). On HI farms, as DIM increased, lying time increased ($P \leq 0.001$), but milk yield and parity did not ($P=0.42$; $P=0.51$). These data suggest that cows in LI systems were more active overall, and lying behavior was more sensitive to circadian rhythms, as well as milk yield, DIM, and parity in comparison to cows on HI farms. Closer examination of the environmental factors influencing variations in lying behavior among LI and HI systems will aid in formulating recommendations to maximize welfare and production on pasture dairies.

### Challenges of assessing welfare in pasture-based dairy production systems

*Robin E. Crossley[1,2], Emer Kennedy[2], Eddie A.M. Bokkers[1], Imke J.M. De Boer[1] and Muireann Conneely[2]*
[1]*Animal Production Systems Group, Wageningen University and Research, 6700 AH Wageningen, the Netherlands,* [2]*Teagasc, Animal & Grassland Research and Innovation Centre, Moorepark, Fermoy, Co. Cork, Ireland; robin.crossley@teagasc.ie*

Pasture-based dairy systems are often perceived to provide cows with the best welfare. Pasture is frequently reported to reduce lameness by providing more comfortable walking and lying surfaces and greater exercise, as well as allowing expression of a wider range of social behaviours. However, pasture-based dairy production is practiced in only select countries, such as Ireland and New Zealand, where weather conditions allow for nearly year-round grass growth. Within these systems there has been little research into the welfare status of the cows. Several methods of assessing animal welfare have been developed and used on-farm, such as the Animal Needs Index in Germany and Austria, the Bristol Welfare Assurance Programme in the UK, and the current gold-standard in Europe for evaluating the welfare of dairy cattle, the Welfare Quality® assessment protocol. However, such assessments were designed to evaluate housed dairy cows and thus do not address the unique challenges of pasture-based systems. Most of these existing assessments aim to include a variety of measures that are animal-based, resource-based and management-based. However, evaluating management and resource measures, such as appropriate housing conditions (e.g. feeding, flooring or bedding type) and how cows function within these conditions, may prove difficult to assess in a pasture-based system, where cows are grazing for the majority of the year. Additionally, because the descriptions of assessment measures are targeted to cows housed indoors, they may be open to interpretation by assessors of grazing cows, leading to inconsistent application of the protocol. Requirements for feed provision, for example, that describe feed availability at a bunk or feed alley may be difficult to translate into a system where adequate feed provision is related to grazing time and residual grazing height. Existing welfare assessments may also exclude evaluation of important contributing factors, such as daily walking distance between paddocks and the milking parlour, which may greatly impact lameness. Additionally, certain included measures, such as avoidance distance, that must be performed under specific conditions may be difficult or impractical to perform with animals at pasture. Based on this review of existing protocols, we suggest the development of a practical and comprehensive welfare assessment targeted towards pasture-based dairy production systems. The assessment should incorporate pasture-specific descriptions and additional measures such as walking distance, roadway condition and rumen fill. Not only would this benefit total grazing systems, but also indoor housing systems in countries that practice seasonal grazing, such as the Netherlands, France, Germany and Belgium. Resulting data would identify areas where pasture-based systems are excelling at maintaining a high level of animal welfare, as well as areas that need improvement, and provide a benchmark for welfare status between farms.

**Interests and values of dairy producers and consumers on animal welfare in Southern Brazil**

*Andreia De Paula Vieira¹, Tatiane V. Camiloti², Raymond Anthony³, Ediane Zanin², Carla Barros² and Jose A. Fregonesi²*
¹CPUP, R. Prof Pedro V.P. de Souza 5300, 81280330, Brazil, ²UEL, Campus Universitário, 86057970, Brazil, ³UAA, 3211 Providence Dr, 99508, USA; apvieirabr@gmail.com

The objectives of this research are to gauge and explore consumer and producer attitudes towards animal welfare and to link the findings to sustainable production methods. Consumers and producers in Brazil are increasingly interested in the intersection of animal welfare, the environment and social sustainability. Understanding central values and aspirations of Brazilian stakeholders on which aspects of animal welfare to address is essential to enhance stakeholder engagement and for determining policies and practices to adopt. The methods include analyzing, documenting and comparing the attitudes of Brazilian dairy consumers and producers towards animal welfare concepts applied to dairy systems of high/low external inputs with a view towards understanding its importance in informing the sustainability of the dairy production chain. Interviews and follow up focus groups gauging producers and consumers' attitudes and preferences towards animal welfare concepts applied to dairy production using the SAFA framework were employed. Target samples were consumers and farmers in Rio Grande do Sul, Santa Catarina and Parana, important dairy states. IBGE distribution data informed sample representativeness. Preliminary survey data from Parana found 69% of consumers prioritized lower product prices over nutritional information (58%), health (47%) and product guarantee for quality and types of packaging (44%). Consumers were primarily concerned about environmental sustainability (86%), reliable labeling (84%) and dairy cow welfare (80%). Consumers varied in their perceptions of animal welfare: 98% held good health, normal growth, absence of diseases or injuries constituted good welfare, followed by absence of negative affective states (pain, discomfort, fear), and having available veterinary services (93%). 84% reported that well-trained producers contributed to better quality of life for cows. The results will guide recommendations for producers to better aggregate their production quality and compete effectively in light of new marketing requirements and evolving consumer demand for dairy products according to animal welfare standards and sustainability expectations; and facilitate future studies to validate local and scientific indicators of animal welfare on farms, at processing industries and in retail stores. The data reveals tools for improving producer-consumer communication about animal welfare and links human and animal quality of life issues to sustainability, economic profitability, environmental protection, technological innovation, and a broader range of socio-cultural aspirations. It will inform: (1) Transformation of welfare models to include consumer concerns and values, opportunities for infrastructure development, and government or corporate incentive programs; (2) Dissemination of reliable information to consumers about the nature of their dairy system and the commodities produced; and (3) Development of more effective channels of communication between actors in the production and distribution chain, as a way to minimize risks for the industry.

## Development of a rising and lying-down ability index in dairy cattle and its relationship with other welfare outcome measures

A. Zambelis, M. Gagnon-Barbin, J. St John and E. Vasseur

McGill University, Department of Animal Science, Macdonald Campus, 21111 Lakeshore Rd., Ste-Anne-de-Bellevue, QC, Canada, H9X 3V9, Canada; elsa.vasseur@mcgill.ca

The objective of this study was to develop two scoring systems to evaluate the rising and lying-down ability of dairy cows, and to assess the relationship of abnormal rising and lying-down behavioral indicators with other welfare outcome measures. Primiparous (n=12) and multiparous (n=35) lactating Holstein cows (133.8±75.2 DIM) were divided between two start dates and video recorded 1 d/wk for 10 consecutive weeks in a tie-stall facility. In week 1 of start 1, all of the abnormal rising and lying-down behaviors were recorded by a single trained observer. Indicators of rising abnormality included rising time (sec), number of rising attempts, and presence (1) or absence (0) of tie-rail contact, backward movement on knees, and delays during rising. Indicators of lying-down abnormality included intention time (sec), lying-down time (sec), number of lying-down attempts, and presence (1) or absence (0) of stall contact, hind-quarter stepping, and slipping during lying-down. Summary parameters were created as additional indicators of rising and lying-down abnormality to classify each event as normal (0) or abnormal (1) based on the presence of at least one abnormal behavior. Averaging scores from 4 daytime events and 2 night-time events was sufficient to yield a strong correlation with full 24 h averages for each indicator of rising and lying-down abnormality (rising: r=0.95; lying-down: r=0.90). Three independent observers used this system to score the daily percentage occurrence of each indicator in weeks 2, 3, 6, 8, and 10 of starts 1 and 2. All indicators were scored with high repeatability among observers (≥0.80 κ) after training. Multivariate mixed models were used to assess the association of each indicator as an outcome of other welfare measures. The models included the fixed effects of injury severity, body size, lying time, parity, lactation stage, and stall lameness status, the random effects of cow within start, and the repeated effects of week. The indicators were positively associated with injury severity and body size, and negatively associated with lying time ($P \leq 0.05$). Presence of a low severity dorsal calcanei injury increased occurrence of tie-rail contact during rising by 17% ($P=0.02$) compared to no injury cows. For every cm increase in cow width, rising time increased by 3.3 s ($P=0.03$) and lying-down stall contact occurrence increased by 1.9% ($P=0.01$). Both scoring systems were highly repeatable tools for assessing the extent of abnormal rising and lying-down behaviors in dairy cows, and were associated with other welfare outcome measures.

## Automatic measurement of rumination, feeding, and resting behavior in group-housed dairy cattle: validation of a novel collar

*Lori N. Grinter and Joao H.C. Costa*

*University of Kentucky, Dairy Science Program, Dept. of Animal and Food Sciences, Lexington, KY 40546-0215, USA; lorigrinter@hotmail.com*

Accuracy of precision dairy technology is important because it is a popular method to measure behavior on commercial dairies and in research. These technologies are useful for continuous, automated behavior records without human intervention. A commercially available behavior monitoring collar that measures rumination, feeding, and resting behavior. The study objective was to compare cow behavior measured by the collar to visual observations. Twenty-four lactating Holstein dairy cows (mean ± standard deviation; 190±94 days in milk) were randomly selected for observation. At least 2 wk prior to observation, cows were fitted with collars for acclimatization. Each cow was observed for 240 min within 1 d (07:00 to 09:00 h, and 19:00 to 21:00 h). One trained person observed all cows to avoid inter observer variability. Rumination, feeding, and resting time (min) recorded by the collar were compared to visual observation by Spearman correlation, regression coefficient of determination and Bland-Altman plots analyses. A Spearman correlation coefficient ($r_s$) of 0.95, 0.85 and 0.91 ($P<0.001$) was found for rumination, feeding and resting time, respectively. Also, in the regression a very high coefficient of determination was found for all behaviors (rumination: $R^2$=0.99, feeding: $R^2$=0.94, resting time: $R^2$=0.98; $P<0.001$). A Bland-Altman plot assessed the differences between the collar and visual observations for rumination, feeding and resting time. Mean differences ± standard deviation (collar – observation) were -6.00±8.21; 20.47±16.98; and -15.51±9.44 min, respectively. The 95% confidence interval of the Bland-Altman plot encompassed 100% of the observations of resting time, and all but one cow's observations for both rumination and feeding time. In conclusion, the collar reported rumination, resting and feeding behaviors with high correlation in group housed, TMR fed, Holstein dairy cattle.

**Use of tail movement to predict calving time in dairy cattle: validation of a calving detection technology in dairy cattle**

*Sarah E. Mac, Carissa M. Truman and Joao H.C. Costa*
*University of Kentucky, Dairy Science Program, Dept. Animal and Food Sciences, Lexington, KY 40546-0215, USA; jhcardosocosta@gmail.com*

Early detection of calving allows the farmer to manage the parturient cow, to be present during calving if necessary and to monitor cases of dystocia in dairy cattle. Dystocia, when not assisted, has the potential to increase calf mortality, decrease milk yield, lower conception rate, and increase uterine disorders. The objective of this study was to evaluate the ability of a commercial precision technology, Moocall, to detect the onset of calving in dairy cows. Data from 73 cows were collected from September 2016 to January 2017 at the University of Kentucky Coldstream Dairy. The calving detection device was attached to the tail 4±3 days (mean ± SD) before expected due date, and video was recorded for behavior analysis. The tail mounted technology is programed to send 2 SMS alerts per calving, one at 2 h and the second at 1 h before calving. Accuracy of the calving device was evaluated by comparing the alert times to the actual time of calving. To better understand the calving detection technology monitoring, tail behavior was monitored and analyzed for frequency and duration of tail lifts two h prior to the first alert (baseline period), the h prior to the first alert, and the h prior to the second alert. A lower one-sided analysis for significance of the difference in means was performed, with the average difference between alert 1 and alert 2 as 150 and 90 min, respectively. The average time interval between the first alert and calving was 107±10 min (mean ± SEM, $P<0.01$) and the average time interval between second alert and calving was 71±10 min ($P<0.01$). Majority of cows gave an acceptable alert with 13 cows calving without an alert. However, about 50% alerts were considered a false alert and 25 cows gave more or one false alert. Video was evaluated for the frequency and duration of tail lifts during the control period, h prior to the first alert, and the h prior to the second alert. Mean frequencies were 3.37, 7.95, and 8.47 (lifts/h), respectively. Mean durations of tail lifts were 55, 124, and 134 seconds, respectively. The calving detection device has the potential to alert farmers approximately two h before calving. The farmer being present during birth can reduce dystocia problems and provide timely delivery of colostrum, improving cow and calf health.

## Innovative cooling strategies for dairy cows

*Alycia Drwencke[1], Matthew Stevens[2], Vinod Narayanan[2], Theresa Pistochini[2], Grazyne Tresoldi[1] and Cassandra Tucker[1]*
[1]*University of California, Davis, Animal Science, One Shield Ave, Davis, CA 95616, USA,* [2]*Western Cooling Efficiency Center, 215 Sage St Ste 100, Davis, CA 95616, USA; amdrwencke@ucdavis.edu*

Producers in the Western U.S. commonly use spray water at the feed bunk and fans in the lying area to mitigate heat stress in dairy cows. Spray water cycles on and off and fans turn on once a pre-set activation air temperature is reached. While this can be an effective method of mitigating heat stress, innovative methods are needed to improve sustainability by reducing water and energy use. Our objective was to evaluate the effectiveness and resource efficiency of 4 cooling treatments on behavioral and physiological responses in dairy cows housed in a free-stall barn. We also measured water and energy use of the treatments. The 4 treatments we tested were: (1) conductive cooling, where cooled water mats were buried under the lying area (Mat; activated at 19 °C); (2) targeted convective cooling where cool air was directed toward the cows through fabric ducts at both the feed bunk and lying areas (Targeted Air; activated at 22 °C); or (3) a combination of spray water and fans described above (Baseline, activated at 22 °C); and (4) spraying half the amount of water as in Baseline and moving the fan to the feed bunk to improve evaporation (Optimized Baseline; activated at 22 °C). In a crossover design, groups of cows averaging (± SD) 34.9±5.3 kg/d of milk (n=8 groups; 4 cows/group) were tested for 3 d/treatment. For ethical reasons, both the Mat and Targeted Air also had spray water beginning at 30 °C. We recorded body temperature, posture, and location within the pen every 3 min 24 h/d, and respiration rates every 30 min daily from 10:00 to 19:00 h when air temperature averaged, respectively, 26.2±2.3 and 33.5±2.9 °C. Pairwise comparisons within a mixed model were used to compare each treatment to the Baseline. Average time spent lying and milk production were not affected by treatment ($P>0.1$). Respiration rates did not differ across treatments overall (58±2 breaths/min), but on an hourly basis, cows on Mat had a significantly higher rate compared to Baseline, at h 10 and 11 ($P<0.03$). Body temperature averaged 38.7±0.2 °C across treatments and was higher when on Mat compared to Baseline at h 10, 11, 20, 21, 22, with a difference of 0.2-0.3 °C ($P<0.04$). Average lying time was 51±2.4%/d across treatments but was 56%/h for the Mats during the hours body temperature was higher, indicating that they were being used during those periods. Taken together, these results indicate that the Mat treatment did not effectively reduce early indicators of heat load, compared to Baseline. In contrast, both Targeted Air and Optimized Baseline were both effective, but differed in other aspects of sustainability. Targeted Air used the least amount of water, but the most energy of all options tested. In conclusion, more efficient heat abatement options can be identified, particularly an Optimized Baseline strategy, which cut water use in half, used the same amount of energy as the Baseline, and maintained similar responses in cows.

## Measuring the effects of short-term overstocking, heat stress, or combination on welfare of lactating dairy cows

*Amanda R. Lee[1], Gina M. Pighetti[1], Rick J. Grant[2], Janice L. Edwards[1] and Peter D. Krawczel[1]*
[1]*University of Tennessee Knoxville, Department of Animal Science, 2506 River Drive, Knoxville, TN 37996, USA,* [2]*William H. Miner Agricultural Research Institute, 1034 Miner Farm Road, Chazy, NY 12921, USA; alee90@vols.utk.edu*

Overstocking (OS) and heat stress (HS) are two primary management challenges that impact dairy cow welfare. The objective was to evaluate the effects of a single stressor, 141% stocking density (OS) or no heat abatement (HS), or concurrent dual stressors (OS and HS; OSHS) on behavior, milk production, and milk components of lactating dairy cows. A 4×4 Latin square was implemented on 64 cows (parity & DIM ± SEM; 1.7±0.1 and 129±8) from July to September 2017. Each 14 d period consisted of 7 d acclimation and 7 d data collection. A subset of cows (n=44) were equipped with accelerometers (IceTags) during each acclimation period to measure lying and standing behavior. Cows were housed in four pens (n=16), with natural ventilation on 2 sides, containing sand bedded freestalls. During period 1, pen temperature humidity index (THI) was collected in a HS and non-HS pen and averaged by d. During periods 2, 3, and 4, all pen THI was averaged by pen/d. Respiration rate (RR) was recorded at 16:00±1 h 4×/weekly to determine cows' heat load. Milk production was manually recorded 2×/d and summed by d. Milk fat, protein, and SCS (TN DHIA Lab) were collected once each period during AM and PM milking and averaged by d. The MIXED procedure evaluated the fixed effects of DIM, parity, treatment, and THI on lying time, bouts, steps/d, milk production, milk components, SCS, and all 2-way interactions, blocked by period. The 1st and 99th percentile were removed to make dependent variables normally distributed. Manual backwards elimination removed non-significant 2-way interactions, with all fixed effects remaining in the model, regardless of significance. Respiration rate was 14 to 16 breaths/min greater among HS and OSHS cows, than non-HS cows ($P<0.001$). Elevated RR suggested cows experienced mild HS during all periods. Lying time of primiparous OSHS cows was 89, 51, and 71 min/d less than no induced stressors (CONT), OS, and HS cows, respectively ($P<0.003$). Overstocked and OSHS cows took 186 and 193 more steps/d than CONT cows ($P<0.002$), suggesting increased standing time among these cows. Heat stressed and OSHS cows produced 1.7±0.8 and 2.1±0.8 kg/d ($P<0.001$) less milk than non-HS cows, but did not differ in milk fat, protein, lactose or SCS. Cows exposed to mild, short duration HS, OS, and OSHS decreased milk production and increased steps/d. However, there were no greater production or behavior effects on cows exposed to short-term concurrent dual stressors vs single stressors.

### Loosening the ties we put on dairy cows

*Véronique Boyer[1], Steve Adam[2], Anne Marie De Passillé[3] and Elsa Vasseur[1]*
*[1]McGill University, Animal Science, 21111 Lakeshore, Sainte-Anne-de-Bellevue, Québec, H9X 3V9, Canada, [2]Valacta, 555 Boul. des Anciens-Combattants, Ste-Anne-de-Bellevue, Québec, H9X 3R4, Canada, [3]University of British Columbia, Dairy Research and Education Center, 6947 Lougheed Highway, Agassiz, British Columbia, V0M 1A0, Canada; veronique.boyer@mail.mcgill.ca*

Although numerous farms in Canada and elsewhere still use tie-stall housing for their dairy cows, very little information pertaining to cow comfort and behavior in such systems is available for producers. The main criticism addressed to the tie-stall system often lies in how it restricts the cow's ability to move, by offering a reduced dynamic space to the animal. The objective of this study was to see whether increasing the length of the tie chain provides cows with an improved opportunity for movement, and to measure how it impacts their rising and lying movements and behaviors. Two treatments were tested: the current recommendation of 1.00 m (control) and a longer chain, of 1.40 m (long). Twenty-four cows (12 per treatment) were blocked by number of parities and stage of lactation, then randomly allocated to a treatment and a stall within one of two rows in the barn for a 10-week period. Leg-mounted accelerometers were used to record lying behaviors, and moments of transitions between lying and standing positions for all cows. The cows were recorded on video for 24 h/week using cameras positioned above the stall. These videos were then used to evaluate the rising and the lying movements of the cows on weeks 1, 2, 3, 6, 8 and 10. Of all the transitions indicated in the accelerometer data, six rising and six lying motions were selected at random. These motions were assessed by a trained observer to detect the presence of abnormal behaviors. Differences between and within treatments over time were analyzed in SAS using a mixed model with treatment, week, and block as fixed effects, and with row and cow as random effects. Data from weeks 1 to 3 were grouped together as the short-term effects, whereas those from weeks 8-10 were grouped together as the long-term effects. Week 6 was used as the mid-term assessment for analysis. Multiple comparisons between terms were accounted for using a Scheffé adjustment. Results indicate that duration of intention movements (exploratory head movements made by the cow prior to lying down) is shorter in cows with longer chains ($13.6 \pm 1.03$ vs $16.8 \pm 1.01$ s for the control, $P=0.048$). It was also significantly shorter in the long term compared to the short-term for both treatments ($13.3 \pm 0.92$ vs $16.9 \pm 0.81$ s, $P<0.05$). Average number of lying bouts per day was numerically, but not significantly, higher in the long chain group ($13.2 \pm 1.09$ vs $12.8 \pm 1.08$ for the control). These results suggest that increasing the chain length improves the cows' ease of movement and transitions, although cows also become more at ease with their surroundings with time. It may provide evidence of a potential way to improve the dynamic space provided to cows in tie-stall systems, using a simple, affordable modification.

## Affiliative and agonistic interactions in dairy cows: a social network approach

*Inès De Freslon[1], Beatriz Martínez-López[2], Jaber Belkhiria[2], Ana Strappini[3] and Gustavo Monti[1]*
*[1]Universidad Austral de Chile, Department of Preventive Veterinary Medicine, Isla T, Valdivia, Chile, [2]University of California-Davis, Center for Animal Disease Modeling and Surveillance, 1 Shields Ave, Davis, USA, [3]Universidad Austral de Chile, Department of Animal Science, Isla T, Valdivia, Chile; imdefreslon@postgrado.uach.cl*

Dairy cattle social structure is regulated by affiliative and agonistic interactions. There is little knowledge on how social interactions can shape the contact structure of cow herds and vice versa. This study aimed to analyze the allogrooming and headbutting contact structure of a group of dairy cows using a network approach. We worked with a group of 38 dairy cows kept in a pasture-based system. Behaviour sampling with all-occurrences recording was used to register the contacts. Cows were observed four hours per day for six weeks. Two contact networks were created for allogrooming and headbutting. Using exponential random graph models, we modeled the probability of contact based on individual and structural effects. We included the following individual attributes: calving number, age, weight, reproductive status, social rank, and time of entrance to the group. The main structural effects tested were reciprocity and geometrically weighted edgewise partners (i.e. transitivity, when two interacting individuals are more likely to have multiple shared partners than expected by chance). Social rank was obtained using the dominance index (DI) of Lamprecht. For each pair, the dominant cow was determined as the one with the highest number of headbutts towards the other (subordinate). The DI of each cow was obtained dividing the number of its subordinates by the number of animals in the group. It was ranked as high (DI≥60), medium (60>DI≥40) and low rank (DI<40). Older cows performed more allogrooming (OR=1.3, $P<0.0001$) and headbutting (OR=1.4, $P<0.0001$) than younger cows. There was significant age homophily for both behaviours, meaning that cows interacted mainly with same-age individuals (allogrooming: OR=1.2, $P<0.001$; headbutting: OR=1.6, $P<0.0001$). Newly entered individuals (cows that entered the group during the observation period, n=16), groomed each other's significantly more than cows present from the beginning of the observations (OR=2.4, $P<0.0001$). They also performed and received more headbutts than the rest of the cows (performed: OR=2.0, $P=0.005$; received: 1.8, $P=0.04$). Medium ranking cows displayed more allogrooming than high ranking cows (OR=2.8, $P<0.0001$). Interestingly, low ranking individuals groomed mainly cows of their same social rank (OR=2.0, $P=0.01$). High ranking cows received significantly more grooming than the cows of other ranks. Cows with confirmed estrus during the observation period groomed considerably more than non-estrus cows (OR=1.7, $P<0.0001$). We observed a high tendency for reciprocal allogrooming (OR=2.3, $P<0.0001$), while headbutting was mainly unidirectional (OR=5.8, $P<0.0001$). There was also a moderate allogrooming transitivity (OR=1.8, $P<0.0001$). This study provides further evidence that social network analysis and its hypothesis testing tools can improve the understanding cows' social behaviour.

### Do bells affect acoustic perception of cows?

*Edna Hillmann[1,2], Julia Johns[1,3] and Antonia Patt[1,4]*
*[1]ETH Zürich, Environmental System Sciences, Institute of Agricultural Sciences, Universitätstr. 2, 8092 Zurich, Switzerland, [2]Humboldt-Universität zu Berlin, Animal Husbandry, Faculty of Life Sciences, Philippstr. 13, 10115 Berlin, Germany, [3] University of Kassel Witzenhausen, Farm Animal Behavior and Husbandry Section, Faculty of Organic Agricultural Science, Nordbahnhofstraße 1a, 37213 Witzenhausen, Germany, [4]Friedrich-Loeffler-Institut, Federal Research Institute for Animal Health, Institute of Animal Welfare and Animal Husbandry, Dörnbergstr. 22-25, 29223 Celle, Germany; edna.hillmann@hu-berlin.de*

In alpine regions, cows are often equipped with bells during pasture season to ensure that farmers can locate them. In an earlier study, we found that bells reduce feeding and lying duration of cows on pasture. Sounds with amplitudes similar to cowbells have been shown to induce avoidance behaviour in cattle. Considering that cows have a well-developed hearing capacity, constant exposure to the chime of a bell may also affect cows' hearing. The aim of the present study was therefore to test the behavioral reactivity of cows that were either equipped with a bell every summer or were not used to bells towards a sound playback of low and high amplitude pink noise in their home pen. Additionally, we tested whether wearing earplugs, mimicking hearing impairment, reduces the cows' reactivity towards the sound. On 24 farms, half of them routinely using cowbells, 96 Brown Swiss cows (3-10 years old) were tested in a 2×2 factorial cross-over design in balanced order. Each cow was individually exposed to 65 and 85 dB, without and with earplugs, in a separated section of the feeding rack. The effect of bell experience, amplitude and earplugs on the latency to a behavioural reaction (change of ear/head-posture, freezing, retreat) to a 5-second playback was analyzed by using linear mixed effects models, considering dependencies within the dataset (trial nested in individual nested in farm). Cows reacted faster without earplugs (0.62 vs 1.47 s, $P<0.001$) and when they were exposed to 85 dB compared with 65 dB (0.65 vs 1.44 s, $P<0.001$). The proportion of cows avoiding the stimulus by leaving the feeding rack after onset of the playback was reduced by bell experience (37 vs 19%, $P=0.027$) and earplugs (37 vs 17%, $P<0.001$), and was increased after the 85 dB compared to the 65 dB stimulus (16 vs 37%, $P<0.001$). However, no indication of complete hearing loss due to bells was found. The 85 dB stimulus triggered faster responses and greater avoidance compared with the 65 dB stimulus, with bell experience and earplugs leading to a decrease in avoidance of the stimulus. This may reflect an altered perception of the sound stimulus due to routine bell exposure.

### Space availability and avoidance distance in cows in shelters (gaushalas) in India

*Arvind Sharma and Clive J.C. Phillips*
*Centre for Animal Welfare & Ethics, School of Veterinary Science, The University of Queensland, Gatton Campus, 4343 Gatton Campus, Australia; arvind.sharma@uqconnect.edu.au*

Gaushalas are traditional cow retirement homes in India where old, infirm, unproductive and abandoned cows are sheltered until they die of natural causes, as cow slaughter is prohibited by law. With limited gaushala capacity, it was hypothesized that space availability would be a welfare issue but that close contact between humans and cattle would be less of an issue, given that Indian farming systems mostly involve regular close contact. Fifty-four cow shelters in 6 states of India were visited in a cross-sectional study and 30 cows in each shelter were randomly selected for assessing avoidance distance. Median herd size in each shelter was 137 cows, there were on average 2 sheds with an area per cow of 2.73 $m^2$. Outside the shed, cows had an average space of 5.9 $m^2$ each in yards. Avoidance distance was assessed 1 h after morning feeding, as prescribed by the Welfare Quality® protocol. Thirty cows standing at the feeding manger were approached from the front at a rate of one step/sec, starting at 2 m from the manger. The distance between the assessor's hand and the cow's head was estimated at the moment the cow moved away and turned its head. The median avoidance distance score was 1.53 with an interquartile range of 0.93, on a scale of 1: touched; 2: 0-50 cm; 3: 51-100 cm; and 4: >100 cm. Approximately 52% of the cows allowed themselves to be touched by the assessor. In shelters where cows were tethered, they had an average space allowance of 5.09 $m^2$/cow. Space availability was therefore limited, compared with modern loose-housed dairy farming systems, but cows were more easily approached. The cows in shelters which provided access to yards had comparatively more contact time with handlers and the avoidance distance was negatively correlated to access to yards in such shelters($r=-0.351$, $P=0.009$).

### Turning off dogs' brains: fear of noises affects problem solving behavior and locomotion in standardized cognitive tests

*Karen L. Overall[1], Arthur E. Dunham[1], Peter Scheifele[2] and Kristine Sondstrom[3]*
*[1]University of Pennsylvania, Biology, 415 University Ave., Philadelphia, PA 19104, USA, [2]University of Cincinnati, 3202 Eden Ave, Cincinnati, OH 45267, USA, [3]University of Akron, 225 S. Main Street, Akron, OH 44325-3001, USA; overall.karen@gmail.com*

As part of a problem-solving study, pet dogs known to react with distress to loud noises were solicited. Based on human studies, we hypothesized that affected dogs would process auditory cognitive stimuli (Auditory Middle Latency Response; AMLR) differently than unaffected dogs. We further hypothesized that unaffected dogs would perform better than those afraid of noise. Dogs were evaluated using a standardized, validated, semi-quantitative objective questionnaire from which an Anxiety Intensity Rank (AIR) score reflecting number of categories of noise reactions, behaviors and intensities was calculated for each dog. Each dog underwent a 13 item problem-solving test (CITP) designed to evaluate all cognitive domains. The final test contained a deliberately mildly provocative noise stimulus. During testing, dogs wore collars containing accelerometers using custom firmware which provided second-by-second 3D movement data. Awake auditory testing was performed on 19 dogs whose AIR scores indicated a distressed reaction to noises and 15 dogs whose AIR scores indicated no distressed reaction, as confirmed by testing. Data were evaluated with parametric and non-parametric tests. AIR scores for the 2 groups differed significantly (Permutation tests, $P<0.0001$), although the affected group was only mildly affected. AMLR did not differ between affected and unaffected dogs (Permutation test; $P>0.4$), but there was a highly significant (Spearman Rank Correlation; $P<0.001$) relationship between auditory brainstem response wave V measurement and noise reactivity. Affected dogs took longer to solve the tasks and, overall, did more poorly (Fisher's exact test; $P<0.05$). The accelerometer data, when analyzed by number of epochs and maximum deviations, revealed that movements of affected dogs during testing were more erratic, less fluid, and subject to both greater extreme deviations and longer pauses than were the movements of unaffected dogs. Finally, a significant number of dogs who reacted to noises (n=6) could not complete the auditory testing (G test; $P<0.0294$). Even dogs mildly affected with fear of noises differed from unaffected dogs, and performed more poorly on problem-solving tests, in part, because their movements are characterized by a high degree of physical and behavioral/emotional reactivity. Reactions to noise affect how these dogs move, which may affect every investigatory and interactive aspect of their lives. Combined AIR scores × movement measures may be used to assess welfare in pet dogs.

## Improving canine welfare in commercial breeding operations: evaluating rehoming candidates

*Judith Stella[1], Traci Shreyer[2], James Ha[3] and Candace Croney[2]*
*[1]USDA-APHIS, Center for Animal Welfare, 625 Harrison St, West Lafayette, IN 47907, USA, [2]Purdue university, Comparative pathobiloby, 625 Harrison St, West Lafayette, IN, 47907, USA, [3]University of Washington, Psychology, P.O. Box 351525, Seattle, WA 98195, USA; judith.l.stella@aphis.usda.gov*

This pilot study aimed to develop criteria for identifying commercial breeding (CB) dogs that may be at greater risk for problems associated with rehoming and to identify breeder practices associated with dogs scored as high vs low risk for problems during transitioning to a new home. Adult dogs (n=283) over 18 months of age from 17 CB kennels located in Indiana and Illinois, USA were assessed. A field instantaneous dog observation (FIDO) tool was used to assess behavior during a 4-step stranger approach test as follows: (1) approach the home pen and stand quietly; (2) offer a treat over the pen door; (3) open the door and offer a treat; (4) attempt to pet the dog. At each step, the dog was scored as either red (fearful), green (affiliative, neutral), or yellow (ambivalent) for a possible total of 34 points. After the behavior assessment, 50 mg of hair was shaved from each dog's lower back for analysis of cortisol concentration (HCC). Each facility owner was interviewed using a questionnaire developed to identify breeder management practices. The risk for problems during rehoming was predicted to increase as dogs' scores on the stranger approach test decreased. To calculate the percentage of dogs at risk for transition problems, the criteria for lower risk was set so that dogs had to score green or yellow at all steps. Of the study population, 41.7% met the criteria for lower risk of transition problems, which ranged across facilities from 10-89% of the dogs. Mean HCC (n=266) was 8.92 pg/mg (2.01-108.99±14.24 pg/mg) while the mean HCC at each facility ranged from 5.65-38.48 pg/mg. The mean dog to caretaker ratio was 19.4/1 (min 7.25, max 34.7). All facilities reported that some type of enrichment was offered; 12/17 reported providing some type of exercise, ranging in frequency from 20 min, 1 day/week to 20 min, 6 days/week; 9/17 provided socialization to puppies, and 7/17 socialized adult dogs. Preliminary results of a principal components analysis found 5 independent dimensions representing management factors. A mixed effects logistic regression model on HCC using dimension factor scores and FIDO score as independent variables and facility as a group variable indicated significant inverse effects of the components related to exercise ($P$=0.007, -1.38, 95% CI -2.38, -0.374) and positive effects of dog to caretaker ratio ($P$=0.03, 2.52, 95% CI 0.264, 4.775) but FIDO score was not significantly associated with HCC ($P$=0.62, 0.488, 95% CI -1.44, 2.42). The results of the stranger approach test suggest that more than half the dogs' assessed exhibited fearful responses and are potentially at greater risk for problems transitioning to a new home. Great variation existed in the management practices employed by the breeders studied with some practices associated with higher HCC. Further research is needed to assess the generalizability of these results in a larger population of dogs and kennels using a multi-stage behavioral assessment.

## The predictive validity of an observational shelter dog behaviour assessment

*Conor Goold and Ruth C. Newberry*
*Norwegian University of Life Sciences, Animal and Aquacultural Sciences, Arboretveien 8,*
*Husdyrfagbygningen, Aas 1430, Norway; c.goold@leeds.ac.uk*

Shelter dog behaviour assessments, such as test batteries conducted at one time point, have been inconsistent in predicting dog behaviour after adoption. We provide the first evaluation of the predictive validity of a longitudinal observational-style assessment. The behavioural responses of 240 dogs while at a shelter were recorded in 7 different contexts on a near-daily basis or as often as the contexts occurred: interactions with: (1) known people (met at least once before); (2) unknown people; (3) dogs; (4) toys; (5) food (e.g. eating from the food bowl), and general behaviour; when (6) inside; and (7) outside the kennel. Post-adoption data on behaviour in the same 7 contexts (with 6) and 7) representing behaviour inside and outside the house, respectively), were gathered by two shelter behaviour specialists via telephone interviews with new owners 2-3 and 5-6 wk post-adoption. Pre- and post-adoption behaviour were recorded using the same ordinal scales indicating increasing behavioural problem severity, comprising 10-16 behaviours depending on the context (e.g. from `Friendly' to `Reacts to people aggressively' when meeting people). Across contexts, the behaviours were categorised into green (no problem for adoption), amber (may need behavioural modification training before adoption/an experienced owner) and red (needs behavioural modification training/ an experienced owner) codes so that behaviour was comparable across contexts. Data were analysed using a joint hierarchical Bayesian ordinal probit model, with pre- and post-adoption behaviour in each context as dependent variables in two distinct regression models linked by correlations between their dog-varying random effects. Estimated parameters pertain to the latent metric scale assumed to generate the green, amber and red ordinal responses. Across the 7 contexts, >90% of codes were green both pre- and post-adoption, with no significant change in the average latent scale behaviour score across dogs between pre- and post-adoption time points (mean difference: 0.20, 90% CI: -0.52, 0.95). However, at the individual level, dogs' average behaviour at the shelter was not significantly correlated with their behaviour 2-3 and 5-6 wk post-adoption (2-3 wk: 0.19, 90% CI: -0.03, 0.42; 5-6 wk: 0.09, 90% CI: -0.29, 0.42). Also, the number of observations of aggression towards people and dogs at the shelter (range: 0 to 27 per dog) was not significantly related to amber- and red-coded behaviour reported across contexts post-adoption (b=0.07; 90% CI: -0.03, 0.18). Across dogs, behaviour was relatively stable between 2-3 and 5-6 wk post-adoption, but the likelihood of amber and red codes for behaviour in the new owner's home was significantly higher at 5-6 than 2-3 wk (b=0.99; 90% CI: 0.54, 1.47). These results indicate that there remains substantial uncertainty in predicting shelter dog behaviour after adoption based on longitudinal observational-style assessment. In practice, however, most dogs' behaviours were categorised as green both pre- and post-adoption meaning that, while observational assessments may not improve predictions of behaviour compared to test batteries, large changes in behaviour on the green, amber and red scale were unlikely.

## Refining on-site canine welfare assessment: evaluating the reliability of field instantaneous dog observation (FIDO) scoring

*Lynda Mugenda, Traci Shreyer and Candace Croney*
*Purdue University, Comparative Pathobiology, 625 Harrison Street, 47907, USA;*
*lmugenda@purdue.edu*

Accurate assessments of behavior and welfare are needed to evaluate the state of domestic dogs maintained in commercial breeding (CB) and other types of kennels. Field assessments of dogs' states of being must be valid, reliable, and efficient. However, observer subjectivity and situational variation in dogs' responses pose a challenge to incorporating behavioral metrics into welfare assessment tools. The published field instantaneous dog observation (FIDO) tool, designed to capture the immediately observable physical and behavioral status of dogs in kennels, was used to determine: (1) reliability of behavioral scoring when used by novice raters; and (2) whether and to what extent dogs' behavioral responses to stranger approach changed during a 30-second observation period. Physical health metrics incorporated into the FIDO tool include dog body and coat condition, cleanliness, and observable evidence of illness or injury, such as sneezing, coughing, or wounds. Dogs' behavioral responses to stranger approach are organized into three categories: red, indicating a fearful response to approach, green, indicating an affiliative or neutral response, and yellow, indicating an ambivalent response. In study one, the behavior assessment component of the FIDO tool was conducted by two novice raters with 50 dogs housed at two US shelters. A stranger approached the home pen of each dog in a non-threatening manner, stood quietly, extended a hand to the dog and scored the response while the test was video-recorded. Intra-reliability was assessed by comparing each rater's live observation scores with their scores of the same dogs using video-recordings. Inter-rater agreement between scores from video-recordings was also calculated. In study two, 81 commercial breeding dogs maintained at four USDA-licensed CB facilities in the US were approached by one researcher and scored once every five seconds for 30 seconds. Of the 81 subject dogs, 59 met the criteria for full sampling. Using Cohen's kappa, raters showed almost perfect agreement between their own scores of live and video-recorded shelter dog responses (kappa=0.83, 0.89) and between each other's video-recorded scores (kappa=0.84), indicating high intra- and inter-rater reliability. Results from study two indicated that over a 30 second time-frame with 5-second increments, 95% of the dogs showed no change in their behavioral response to approach. This suggests that the first 5 seconds of scoring provide a reliable time-point for assessing behavior using the FIDO tool and indicates no benefit to extending the FIDO scoring period to gauge dogs' immediate responses to stranger approach. These results provide evidence of reliability of the behavioral component of the FIDO tool even when used by trained novices. Further research is needed to validate scores obtained using the tool against other indicators of immediate welfare state and against long-term indicators of overall welfare.

## Dimensions of personality in working dogs: an analysis of collected data on future guide and assistance dogs over 37 years

P. Plusquellec[1], N. Dollion[1,2], A. Paulus[1], G. Goulet[1,2], M. Trudel[2], N. Champagne[2], N. Saint-Pierre[2] and E. Saint-Pierre[2]
[1]Laboratoire d'Obervation et d'Éthologie Humaine du Québec, École de Psychoéducation, Université de Montréal, Pavillon Marie-Victorin, 90 avenue Vincent-d'Indy, H2V 2S9, Montréal, Canada, [2]Fondation Mira, 1820 Rang NO, QC J0H 1S0, Sainte-Madeleine, Canada; dollionnicolas@gmail.com

Defining and predicting a dog's temperament is of major concern for groups providing assistance and/or working dogs. The Mira Foundation is a non-profit organization offering assistance dogs to autistic children and to individuals with motor or visual disabilities. Since its beginning in 1981, the foundation has registered behavioral data on its dogs relying on rigorous and standardized procedures. First, at 6 and 12-month-old, while dogs are in a foster family, data is collected with questionnaires assessing the presence or absence of specific behaviors (e.g. problematic behavior, excitement or fear behavior, attitude toward humans and animals). Then, at 12 months of age, just before getting proper training, dog trainers conduct a set of short behavioral tests evaluating dogs on various traits that are key to training (e.g. dominance, sensitivity, confidence). Over the years, Mira has donated over 5,500 dogs (i.e. Bernese Mountain, Labernese, Labrador, Saint-Pierre and some other breeds), and has computerized their evaluations. Exploiting this unique database, this study aimed at investigating four aspects: (1) behavioral dimensions assessing Mira dogs' personality at 6 and 12-month-old; (2) stability of behavioral dimensions across development; (3) impact of breeds on these dimensions; (4) reliability of data collected through questionnaires or behavioral testing. After cleaning up the initial database from missing dog information and measurements, the final sample included 1,757 and 2,752 dogs with a complete foster family questionnaire at 6 and 12-month-old respectively, and 1,965 12-month-old dogs evaluated by a trainer with behavioral testing. Multiple correspondence analysis revealed distinct behavioral dimensions that might characterize dog personality at 6 and 12 months old (i.e. activity, fear/reactivity, responsiveness to training, sociability, submissiveness). Furthermore, some behavioral dimensions observed at 6 months of age were significantly correlated with those found from behavioral testing at 12 months. For example, scores on the fear/reactivity dimension at 6-month-old correlated positively to scores on the submissiveness dimension at 12 months ($r_{695}$=.132, $P<0.001$). Moreover, using analysis of variance, we observed that, according to their breed, dogs did not display the same scores on behavioral dimensions of personality (e.g. activity, $F_{4, 1956}$=18.74, $P<0.001$). Indeed, Bernese Mountain, Saint-Pierre and Labernese breeds displayed results that would be more compatible with assistance or guide dog training. Finally, correlation analysis performed on behavioral dimensions extracted from questionnaires at 12 months and from behavioral tests before training showed a significant correlation between similar dimensions (fear/reactivity, $r_{1965}$=.123, $P<0.001$), thus confirming, to some extent, the reliability of this type of measurement. These results are likely to help Mira in developing and improving their methods to train service dogs.

**The effects of aversive- and reward- based training methods on companion dog welfare**

*Ana Catarina Vieira De Castro[1,2], Stefania Pastur[2,3,4], Liliana De Carvalho E. Sousa[1] and I. Anna S. Olsson[2]*

[1]*ICBAS, Instituto de Ciências Biomédicas Abel Salazar, Universidade do Porto, Ciências de Comportamento, Rua de Jorge Viterbo Ferreira 228, 4050-313 Porto, Portugal,* [2]*i3S, Instituto de Investigação e Inovação em Saúde, Universidade do Porto, Laboratory Animal Science, Rua Alfredo Allen 208, 4200-135 Porto, Portugal,* [3]*National Institute of Biology, Ljubljana, n/a, Slovenia,* [4]*Università degli studi di Trieste, Trieste, Italy; olsson@ibmc.up.pt*

Over time, dogs have become ubiquitous in human households as companion animals. As human expectations of companion dog behaviour have changed, dog training schools offer a means to teach dogs acceptable behaviors in a family context. These schools use a variety of operant conditioning techniques, ranging from reward-based methods relying on positive reinforcement to aversive-based methods relying on negative reinforcement and positive punishment. Although the use of aversive methods has been strongly criticized for negatively affecting dog welfare, these claims do not find support in solid scientific evidence. Companion dog-focused research as well as research on the entire range of aversive-based techniques (beyond e-collars), objective measures of welfare, and robust sample sizes are lacking from previous research. We present the first results of an ongoing study aiming to measure the potential behavioral impact of aversive- and reward-based training methods on the welfare of companion dogs. Dogs are recruited from seven dog training schools in the metro area of Porto (3 reward-based and 4 aversive-based schools) and each dog is video-recorded during three training sessions early in their training programs. The video recordings are analyzed using a continuous sample of the frequency of stress-related behaviors (e.g. lip lick, body shake, yawn, and vocalizations – bark, whine, yelp, growl) and using an instantaneous scan sample technique 20 s for overall behavioral state (tense, relaxed, low, excited) and panting. Our results based on the 50 dogs analyzed so far showed that dogs undergoing training at aversive-based training schools displayed a significantly higher frequency of stress-related behaviors than dogs undergoing training at reward-based schools ($t=-5.62$, $P<0.001$). Moreover, within aversive-based schools, we found that the higher the frequency of aversive stimuli used, the higher the frequency of stress-related behaviors displayed by the dogs ($F=11.72$, $P<0.001$). We also found significant differences in the behavioral state displayed by the dogs, with dogs undergoing training at aversive-based schools spending more time in a tense ($t=-8.168$, $P<0.001$) or low state ($t=-2.391$, $P<0.001$) and dogs undergoing training at reward-based schools spending more time in a relaxed state ($t=-4.855$, $P<0.001$). Finally, dogs undergoing training at aversive-based schools panted significantly more than dogs undergoing training at reward-based schools ($t=-6.748$, $P<0.001$). Together, these results suggest that dog welfare may be at stake during training sessions in aversive-based schools. To understand whether training method affects dogs outside the training situation, we are measuring cognitive bias and owner-dog relations.

### Service dogs for epileptic people, a systematic literature review

*Amélie Catala, Hugo Cousillas, Martine Hausberger and Marine Grandgeorge*
*University of Rennes 1, Laboratoire EthoS, Ethologie Animale et Humaine, Station Biologique de Paimpont, 35380 Paimpont, France, Metropolitan; amelie.catala@univ-rennes1.fr*

Recently, there has been a rising interest in service dogs for epileptic people. Indeed, there are reports of dogs that are spontaneously sensitive to epileptic events of their owners. Two types of service dog are considered; dogs can be trained to perform, or spontaneously demonstrate, both roles. First, the seizure-alert dogs which seem to anticipate seizures and warn their owner accordingly. Second, the seizure response dogs which demonstrate specific behaviors during or immediately after a seizure. However, there is a lack of rigorous clinical trials to confirm these abilities and know more about the mechanisms potentially involved. The purpose of this review is to present a comprehensive overview of the validated scientific research on seizure-alert/ response dogs for epileptic people. The specific goals are to: (1) identify the existing scientific literature on the topic, and then describe; (2) the characteristics of seizure-alert/response dogs; (3) evaluate the state of evidence base; and (4) the reported outcomes of seizure-alert/ response dogs. To perform a systematic review, we followed the preferred reporting items for systematic reviews and meta-analyses (PRISMA) guidelines of the empirical literature. Out of 28 studies published in peer-reviewed journals dealing with dogs for epileptic persons, only 5 were qualified for inclusion. Four were self-reported questionnaires, the last one being the only prospective study in the field. Reported times of alert before seizure varied widely among dogs (with a range from 10 seconds to 5 hours) but seemed to be reliable (accuracy from ≥70 to 85% according to owner reports). Alerting behaviors were generally described as attention-getting. The alert applied to complex partial and generalized tonic-clonic seizures as well as atonic and absence seizure. Pet dogs as dogs recruited as seizure-alert dogs generally varied in size and breed. Training methods differed widely between service animal programs, partially relying on hypothesized cues used by dogs (e.g. variations in behavior, scent, heart rate, etc.). However, none of the studies included a comparison condition, and only one looked at the treatment condition, using a pre-post design and a follow-up assessment. There was a low level of methodological rigor in most studies (e.g. regarding medical diagnosis, self-reporting bias, choice of study design, etc.), indicating the preliminary nature of this area of investigation. Furthermore, most studies indicated an increase in quality of life when living with a dog demonstrating seizure-related behavior. In addition, owner reports suggest that the presence of a dog with seizure-related behavior had an impact on epilepsy, reducing the seizure frequency of epileptic persons. In conclusion, scientific data are still too scarce and preliminary to reach any definitive conclusion as to the success of dogs in alerting that a seizure will come, the cues on which this ability may be based, the type of dog that could be most successful and the type of training that should be done. While these preliminary data suggest that this is a promising topic, further research is needed.

## Stranger-directed aggression in pet dogs: owner, environment, training and dog-associated risk factors

*Hannah Flint[1], Jason Coe[1], David Pearl[1], James Serpell[2] and Lee Niel[1]*
*[1]University of Guelph, Population Medicine, 50 Stone Rd E, Guelph, ON, N1G 2W1, Canada,*
*[2]University of Pennsylvania, Clinical Studies, 3800 Spruce St, Philadelphia, PA 19104, USA;*
*flinth@uoguelph.ca*

Stranger-directed aggression (SDA) in dogs can result in human injuries, impair the human-animal bond, and lead to an increased risk of physical punishment, relinquishment or euthanasia. Our objective was to determine risk factors for dogs that display SDA using the previously validated, owner-completed canine behavioural assessment and research questionnaire (C-BARQ), with additional questions added relating to dog impulsivity (dog impulsivity assessment scale; DIAS), demographics, training, and environment, and owner demographics and personality (ten item personality inventory; TIPI). The online questionnaire was distributed to current dog owners through snowball sampling via social media. Data were analyzed using a mixed linear regression model with household modeled as a random intercept (n=2,760 dogs; 2,255 households). Dogs had higher SDA scores if they had mild ($P<0.001$) or severe ($P<0.001$) fear of strangers (relative to no fear), or had higher DIAS impulsivity scores ($P<0.001$). SDA scores were also higher for dogs that were male ($P<0.001$), neutered for behavioral reasons ($P<0.001$), or fed on a schedule ($P=0.036$). Dogs had lower SDA scores if crated when left alone ($P<0.001$), and exercised on leash ($P=0.004$), off leash ($P<0.001$) or at the dog park ($P=0.010$). In addition, breed group was associated with SDA, with hounds scoring lower than herding breeds ($P<0.001$). When looking at the effect of early experiences, dogs exposed to strangers less than once per month as a puppy ($P<0.001$), and dogs with a history of abuse ($P<0.001$) had higher SDA scores. In addition, dogs that were scared ($P<0.001$) or indifferent ($P=0.006$) towards strangers as a puppy had higher scores compared to dogs that were excited as puppies. When looking at training-related factors, SDA scores were higher when owners used head halters ($P<0.001$), no-pull harnesses ($P=0.002$), shock collars ($P=0.001$), or choke chains ($P=0.017$), and if the owners used physical punishment ($P<0.001$), or avoided situations where the dogs performed undesirable behaviours ($P=0.024$). However, scores were lower if the owners asked for a different desirable behaviour when their dogs performed an undesirable behaviour ($P=0.001$). Finally, when looking at the effect of owner, dogs with owners who rated themselves as extroverted on the TIPI ($P=0.002$) or were not able to correctly identify the absence of aggression in videos ($P=0.007$) had higher SDA scores. These results indicate that the primary risk-factors for SDA relate to dog and training variables, and suggest that it might be possible to reduce SDA through owner education programs focused on appropriate methods for dog socialization, management and training. Association does not necessarily indicate causation, but the findings highlight important areas for future research in longitudinal studies with the aim of identifying dogs at risk, and implementing appropriate training plans to prevent SDA.

### The effect of the Danish dangerous dog act on the level of dog aggression in Denmark

*Björn Forkman and Iben C. Meyer*
*University of Copenhagen, Dept. Vet Anim Sci, Grønnegårdsvej 8, 1870 Frederiksberg C, Denmark;*
*bjf@sund.ku.dk*

In 2010 Denmark introduced a dangerous dog act leading to the ban of 13 dog breeds, mainly of two different types; so called fighting dogs and a number of breeds of guarding herd dogs. Dogs of the banned breeds were not euthanized but were required to wear a leash and a muzzle when in public. Three years after the introduction of the ban we assessed the effect of the ban by using material from a number of sources: (1) a university hospital (Odense); (2) three different major veterinary hospitals in various parts of the country; (3) data from an insurance company; and (4) a representative survey of Danish citizens (both dog owners and non-dog owners; n=4,030). The first three sources used retrospective data to compare the situation from 2008 and 2009, to that of 2011 and 2012. The dog ban had no effect on the incidence or severity of bites towards humans. (The number of people treated for dog-bites at the hospital was 370 prior to and 372 after introducing the dog ban.) The main argument for the dog ban was that people should feel secure in public spaces (where the dogs now were required to be muzzled), no significant difference in the proportion of bites to humans in private vs public spaces was found however (Fisher's exact test: $X^2=2.71$, df=1, $P=0.11$), nor was the proportion of unprovoked attacks, as described by the patients, different for the two time periods (13 vs 10%; Fisher's exact test: $X^2=1.34$, df=1, $P=0.29$). As to bites towards dogs the results were mixed with the Copenhagen veterinary hospital reporting a significant decrease in the proportion of dog bites out of all treated animals (Fisher's Exact: $X^2=1.,35$, df=1, $P<0.0001$), but not in the proportion of bites resulting in severe damages (expected from bites by fighting dogs; Fisher's exact test: $X^2=0.99$, df=2, $P=0.62$). No decrease was found for the other two veterinary hospitals (Fisher's exact: $X^2=0.32$, df=1, $P=0.59$; $X^2=2.78$, df=1, $P=0.11$), nor for the insurance company (Fisher's exact test: $X^2=2.81$, df=1, $P=0.1$). The data from the survey found no decrease, but rather an increase, in the number of bites towards humans (48 persons before compared to 76 after) or dogs (152 before compared to 330 after). It may be that dog owners had forgotten incidents from 2008-2009 but when asked for aggressive interactions that required their dog to be taken to a veterinarian the same pattern emerged (27 instances before vs 45 after). When asked about their opinion of the Dangerous dog act the respondents were overall positive to the ban (62 vs 20%). In conclusion the Danish dangerous dog act did not have the desired effect of decreasing dog aggression directed towards either humans or dogs, but was still well regarded by the general public.

## The effect of public cat toilet provision on elimination habits of ownerless free-roaming cats in old-town Onomichi, Japan

*Aira Seo and Hajime Tanida*

*Hiroshima University, Graduate School of Biosphere Science, 1-4-4, Kagamiyama, Higashi-Hiroshima, 7398528, Higashi-Hiroshima, Japan; airaseosan@hiroshima-u.ac.jp*

Free-roaming cat populations are growing to high densities in residential and tourist areas in old-town Onomichi, Japan, where there is a high concentration of numerous historic temples. The city receives many complaints from residents about cat feces soiling the paths, grass, and air quality of the neighborhood. Additionally, cat feces can contain bacteria, viruses, and parasites that infect humans and pets. The cats also irritate the monks and groundskeepers of temples because the cats use the sites to rest and defecate. The Hiroshima prefectural animal shelter financially supports residents to control the number of ownerless free-roaming cats using Trap-Neuter-Return (TNR), a program in which feral cats are trapped, spayed or neutered, and then returned to their original territory, where a caretaker provides regular food and shelter. However, there are no effective control measures for the neutered cats' elimination habits after they are returned to their original territory. We examined how the provision of public cat toilets affected these habits. The study was conducted in the uptown area of old-town Onomichi at a temple and graveyard where cats regularly defecate. Cat feces were collected and weighed 1×/wk for 4 wk at 5 popular defecating spots on the temple precincts and graveyard to assess the damage caused by the cats. Then, commercial cat repellents were placed at the 5 popular defecating spots plus another 6 spots where the temple wanted to prevent cat elimination. Meanwhile, cat toilets, created by filling repurposed plastic planters with cat litter, were placed at 6 spots where the temple allowed the cats to eliminate. The feces in the 6 toilets and at the 5 defecating spots were collected and weighed every week for 14 weeks. The behavior of cats around the toilets and defecating spots were recorded with 4 trail cameras (Ltl-Acorn 6210MC, 6310MC: the cameras trigger with movement and take videos for 60 seconds automatically). The average weight of feces at the 5 defecating spots was 47.0 g/spot/week before the placement of the toilets and repellents. Seven adult cats were responsible for the feces. However, the feces decreased gradually and significantly after the placement of the toilets and repellents and reached 0 g in the final week of the experiment ($P<0.05$: Kruskal Wallis Test with Shirley-Williams Multiple Comparisons). In contrast, the average weight of feces in the 6 toilets (g/toilet/week) increased gradually and significantly after the placement and reached 65.7 g in the 14th wk. ($P<0.05$: Kruskal Wallis Test with Shirley-Williams Multiple Comparisons). Of the 17 cats, 7 used the toilets. The results showed that provision of toilets and use of repellents in favored locations is effective at changing the elimination habits of ownerless free-roaming cats.

## How dependent are ownerless free-roaming cats on water supplied by voluntary cat caretakers in old-town Onomichi, Japan?

*Hajime Tanida and Aira Seo*

*Hiroshima University, Graduate School of Biosphere Science, 1-4-4 Kagamiyama, Higashi-Hiroshima, 739-8528 Higashi-Hiroshima, Japan; htanida@hiroshima-u.ac.jp*

Our previous research in old-town Onomichi, Japan, counted 204 free-roaming cats, most of which died from illness or injury within a few years. Voluntary cat caretakers regularly fed them, but rarely supplied water at feeding sites. The 5 freedoms of animal welfare stipulate that animals should be free from thirst by having ready access to fresh water. Our objective was to investigate the extent to which the ownerless cats depend on water supplied by volunteers. The study was conducted at a park (1,561 m²) in the old-town. The caretakers started to supply water in addition to the food at the park one year ago. The drinking behavior of cats at the water container was recorded with a trail camera, which was motion activated to record 60 sec videos, for 48 days (4 non-consecutive days/month, 24 h a day) during 12 months from the start of water supply. Additionally, their behavior at the container was recorded continuously for one full week in the beginning and in the middle of the period. Individual cats were identified by their appearances. Throughout the entire uptown area, at the center of which lies the park, route censuses were also conducted 4 times a month during the same period to study the home range of the cats recorded on the camera. A single observer walked the entire route from 10:00 to 16:00 (2.61 km, 1.5 h). When the observer encountered a cat on the route, she photographed him/her and recorded several items of information on a cat identification card. In total, 33 cats were observed at the park during the year. Twenty-six of them visited the container and 24 of them drank the water at least more than 2 times during the period. Furthermore, 4 of the 24 were regular visitors (appeared 60% or more of the 48 obs. days) that formed a group around the water container. There were no other water supply spots around the park according to our censuses. The average number of drinking events per regular visitor per day was 1.64±0.44. The average number of drinking events per day was significantly ($P<0.01$) different among the regular visitors. Three of the 4 regular visitors visited the station to drink the water 7 days in a row in the full week of observations. Their home range was limited to the area around the park based on our route censuses. The average number of drinking events for all cats during the daytime (06:00 to 18:00) was 0.90±0.20, which was significantly ($P<0.01$) more than the average (0.15±0.03) during the nighttime (18:00 to 06:00). The average number of drinking events per day in summer (Jul. to Sep.), which is dry season, was 0.91±0.10, which was more than the average (0.66±0.17) in winter (Nov. to Jan.) but this was not significantly different. Our results show that only a few regular visitors could get access to fresh water and other individuals in the park were probably dependent on dirty water such as puddles of rainwater and gutters, which may increase their risk of exposure to waterborne pathogens and parasites.

## Examining relationships between cat owner attitudes and cat owner reported management behaviour

*Grahame Coleman[1], Emily McLeod[2], Tiffani Howell[1] and Lauren Hemsworth[1]*
[1]*Animal Welfare Science Centre, The University of Melbourne, North Melbourne VIC, 3051, Australia,* [2]*Zoos Victoria, Parkville VIC, 3052, Australia; grahame.coleman@unimelb.edu.au*

Roaming pet cats pose a significant threat to wildlife through predation, and are more likely to be injured than cats which live indoors. Typically, roaming cats have much shorter lifespans than indoor cats. Nonetheless, it is common in Australia for cat owners to permit their cats to roam, and the perception that cats 'need' to spend time outdoors appears to be a deep-seated social norm. An important determinant of domestic animal management and the ensuing welfare outcomes is the quality of the human-animal relationship (HAR). Furthermore, the most direct influence on intended human behaviour is the attitude an individual possesses towards performing the behaviour in question. Azjen's Theory of Planned Behaviour (TPB) is able to both predict and explain motivational influences on human behaviour and to identify target strategies for changing behaviour. The present study is part of a larger project examining the effectiveness of a zoo conservation campaign in changing cat owners' management behaviours when delivered in settings with varying levels of interaction with a zoo animal. This study aims to examine relationships between cat owner attitudes towards managing their cats and their reported cat management behaviours. Forty Victorian cat owners, whose cats spent at least some time outdoors, were recruited, via an advertisement posted on a range of social media sites, to complete an online survey. The survey consisted of a questionnaire concerning owner attitudes towards cats, cat management behaviours and conservation, and a cat management diary where owners reported their daily cat management behaviours. The questionnaire was used to collect information on cats owned, demographic information, cat confinement, cat management behaviours, attitudes to keeping cats confined and attitudes to wildlife conservation. Pearson Product Moment correlations were used to examine relationships between cat owner attitudes towards cat management and cat owner-reported cat management behaviour. Positive attitudes towards having the time to provide the cat with mental stimulation ($r=-0.44$, $P<0.05$), walking the cat on a leash outdoors ($r=0.46$, $P<0.01$), keeping the cat indoors at night for safety reasons ($r=-0.60$, $P<0.01$), keeping the cat indoors at night for conservation reasons ($r=-0.60$, $P<0.01$), and donating to native fauna conservation ($r=-0.38$, $P<0.05$) were all significantly correlated with cats spending less time outdoors (cat owner reported behaviour). A negative attitude towards cleaning out the litter tray was correlated to the cat spending more time outdoors ($r=0.34$, $P<0.05$). These relationships between cat owner attitudes and cat owner reported cat management behaviour are consistent with the TPB. However further research to provide evidence of causal relationships is necessary before the effectiveness of an education campaign to change cat owner management behaviour can be investigated.

**Early evaluation of suitability of guide dogs – behavioral assessment by puppy test**

*Xun Zhang[1], Fang Han[2], Jingyu Wang[3], Katsuji Uetake[1] and Toshio Tanaka[1]*
*[1]Azabu University, Graduate School of Veterinary Science, 1-17-71 Fuchinobe, Chuo-ku, Sagamihara-shi, Kanagawa, 252-5201, Japan, [2]China Guide Dog Training Centre of Dalian, No.9, West Section, Lvshun South Road, Lvshunkou District, Dalian, 116044, China, P.R., [3]Dalian Medical University, Laboratory Animal Center, No.9, West Section, Lvshun South Road, Dalian, 116044, China, P.R.; da1602@azabu-u.ac.jp*

Guide dogs are trained to assist blind and visually impaired people around various difficulties in life. Considering the duration and cost of training a successful guide dog, an understanding of early behavioral suitability of potential guide dogs would be useful for guide dog organizations. Studies on puppy tests have questioned their reliability mainly because of its lack of a scientific base. However, other studies have reported differences within breeds in behavioral profiles between Europe and America, and an effect of regional culture and gene pool. Additionally, the behavioral trait of a breed can change over time. The aim of this study was to determine if early identification of characteristics of dogs best suited for China guide dog work could be done through a puppy test. A total of 19 Labrador retrievers (11 males, 8 females) from the China Guide Dog Training Centre were enrolled in the study. The puppy test used was based on the Japan Guide Dog Association test, adapted from the Cambell's puppy test. The dogs were subjected to the test three times: 1. at 1.5-3.5 months old (first test), 2. at 7-8 months old (second test), and 3. at 13-15 months old (third test). The test included the following thirteen sections: "Fear", "Exploration of the Environment", "Exploration of Humans", "Social Attraction", "Social Dominance", "Following", "Restraint Dominance", "Social Attraction After Constraint (restraint)", "Social Dominance after Constraint", "Noise Reaction", "Chase Instinct (towel and long stuffed toy)", and "Response to Sudden Opening of Umbrella". Responses to each section of the test were video recorded, and assessed and scored on a scale of 0-4. The mean scores of the first, second, and third tests were compared. Consistency of behaviors (the difference of judged raw scores was less than 1) in the three puppy tests was analyzed by Wilcoxon signed-rank test. The number of dogs showing behaviors on the first test consistent with those in the third test (7.54±2.02) was lower than for the comparison between the second test and the third test (11.08±1.94; $P<0.05$). Results showed that the behaviors of the dogs on the second test were more similar to that in the third test compared with that on the first test. However, when comparing specific sections of the tests, we noted that the results of six sections on the first test were similar to those on the third test. These were "Social Dominance", "Following", "Restraint Dominance", "Social Dominance after Constraint", and "Chase Instinct (towel and long stuffed toy)". Thus, these sections of the test may be useful to select the potentially best-suited dogs for guide dog work when the dogs are 1.5-3.5 months old. We will assess whether these tests accurately predict actual outcomes for successful guide dogs. However, further research on large sample sizes is needed.

## Can children of different ages recognize dog communication signals in different situations?

*Helena Chaloupkova, Petra Eretova, Ivona Svobodova and Marcela Hefferova*
*Czech University of Life Sciences, Husbandry and Ethology of Animals, Kamycka 129 Prague, 16500, Czech Republic; chaloupkovah@af.czu.cz*

Adults can recognize contextual, motivational and emotional content of dog vocal communication and are able to assign the different types of calls to the right situation, regardless of the extent and intensity of their experience with dogs. The aim of the study was to find whether age and ownership of pet dogs in the households affect the recognition of dog vocalizations in children. We tested 265 children from kindergarten and elementary schools aged 4-12 (4-5 years: pre-schoolers, 6-7 years: early-school students and 8-12 years: school students). The children completed questionnaires after being presented with audio and video recordings of dog's vocalizing in different situations illustrating three emotional states: anger, joy and sadness. The study was implemented with the approval of the directors of all the participating schools and the parents of all children. Data were analyzed using the logistic regression in the SAS program, PROC Glimmix. We found that the youngest group of children (4-5 year) had the lowest probability of recognizing the audio recordings in comparison with the early school and school students ($F_{2, 322}$=14.39; $P$=0.0001), but no significant differences were found between the other two age groups (early-school and school students). Also, there was a significant difference in the probability of recognition of the three emotional states ($F_{2, 322}$=8.19; $P$=0.0003): 'anger' had a higher probability of recognition than 'joy' or 'sadness'. Age was also found to have a significant effect on the student's ability to recognize dog's vocalization from the video recordings ($F_{2, 149}$=7.96; $P$=0.0005). The pre-schoolers had the lowest probability of correct recognition in all situations in comparison with the school students. A difference between early-school and school students was found only as a trend. The ownership of a dog in the household had no effect in the analyzed models on the children's' ability to recognize these repertoires. Although they may have had direct experience with dogs in the household, the results showed that children under six years of age had only a small likelihood of correctly recognizing dog communication, both audio and visual signals.

**Simplification of electrocardiogram waveform measurement in pet dogs**

*Megumi Fukuzawa[1], Takayasu Kato[2], Tomio Sako[2], Takahiro Okamoto[2], Ryo Futashima[2] and Osamu Kai[1]*
[1]*Nihon University, College of Bioresource Sciences, Kameino 1866, Fujisawa-shi, Kanagawa, 252-0880, Japan,* [2]*NOK Corporation, Tsujido-shinmachi 4-3-1, Fujisawa-shi, Kanagawa, 251-0042, Japan; fukuzawa.megumi@nihon-u.ac.jp*

There have been many reports of the use of human chest-band-type heart-rate monitors for simple and noninvasive heart-rate measurement in dogs. However, measurement using these types of metal electrodes causes displacement of the sensor due to shifts in the dog's posture or walking style; moreover, the resulting data are affected by the individual dog's skeletal structure and coat type. To develop a measuring device that is relatively unaffected by the dog's coat, walking style, or posture, we attempted electrocardiogram (ECG) waveform recording by using rubber electrodes for biomedical signals. If dogs feel uncomfortable in ECG measurements and can reduce its discomfort by improving the equipment, it is considered that the relevance between behaviour and physiological mechanism can be evaluated more clearly. Five healthy pet dogs (three female, two male; aged 19 to 42 months, weight 20.0 to 26.2 kg) participated in the test. Subjects did not receive pretreatments such as shaving. Each dog was held in a supine position with no anesthesia, and three rubber electrodes (positive, negative, and ground electrode) for biomedical signals (connected to a BITalino biomedical recording system; Lisbon, Portugal) were attached to the dog's chest (right and left intercostal and costal-cartilage junction). Lead II was used to derive a reference point for the unipolar chest leads. ECG waveforms were recorded continuously with Bitadroid. The recording range was $\pm 1.5$ mV and the sampling frequency was 100 Hz. After confirmation of stable waveform recording, the dog's posture was shifted to standing, and then to sternal recumbency followed by sitting, for 3 min each. The waveform was recorded continuously in each posture. Recorded numerical values were captured in an Excel file, and the R-R interval and heart rate were calculated. This study was approved by the Nihon University Animal Care and Use Committee (EXC17B001). ECG waveforms could be recorded in all postures in any dog, and there were no disruptions in the waveform when the dog shifted its body position. There was no significant difference in R-R interval ('standing' 0.75±0.19, 'sitting' 0.71±0.11, 'sternal recumbency' 0.77±0.24) and bpm ('standing' 89.74±23.91, 'sitting' 88.92±12.90, 'sternal recumbency' 84.59±22.81) recorded among postures. On the other hand, individual differences were observed on 'standing' and 'sitting' (Steel-Dwass test $P<0.05$, respectively). These results indicate that it is not only possible, but also useful, to record dog ECG waveforms, along with the R-R interval and heart rate, by using rubber electrodes for biomedical signals.

## Precision phenotyping as a tool to automatically monitor health and welfare of individual animals housed in groups

T. Bas Rodenburg[1,2]

[1]Wageningen University, Adaptation Physiology Group, P.O. Box 338, 6700 AH Wageningen, the Netherlands, [2]Utrecht University, Department of Animals in Science and Society, P.O. Box 80.166, 3508 TD Utrecht, the Netherlands; bas.rodenburg@wur.nl

Farm animals are increasingly housed in large group housing systems. Monitoring health and welfare in these large groups can be challenging. In current welfare assessment schemes, attention for animal-based welfare indicators is increasing, resulting in a shift in focus from environment-based to animal-based indicators. However, in group settings monitoring these animal-based welfare indicators is challenging, especially at the level of the individual animal. Focusing on the individual level is relevant in many different settings, e.g. for precision feeding and management, or targeted veterinary care. Also, in the context of animal breeding, focus on the individual level is pivotal; this is frequently done by housing animals of potential interest for selection individually to measure (or evaluate) their phenotypes. In this way, however, no information about the performance of the individual in a group setting can be included in the breeding program, while social interactions can have profound effects on group performance. New genetic methodology now allows the modelling of these social interactions using direct and indirect genetic effects models. This type of methodology can for instance provide information on the propensity of an animal to become a victim of damaging behaviour (direct genetic effect), but also on the propensity to perform damaging behaviour (indirect genetic effect). To utilise this methodology, the ability to measure accurate individual phenotypes in a group setting becomes very important. Breeding companies currently have a strong interest in developing methods for precision phenotyping, relying on new technology that enables tracking of individuals using combinations of different sensors. This should allow the use of group housing in future breeding operations and should allow more accurate phenotyping. In the PhenoLab project, we investigated possibilities for tracking of location, activity and proximity of individual laying hens. To meet that aim, we tracked individual hens during a five-minute Open Field test using two different tracking systems: Ultra-wideband tracking using TrackLab and automatic video tracking using EthoVision. Ultra-wideband tracking consists of an active RFID tag that is placed on the bird in a backpack. This tag is then located by triangulation by four beacons. Comparing distance moved between TrackLab and Ethovision yielded 96% similar results (sample of 24 hens). In a second step, the ultra-wideband tracking was also used to measure differences in activity between high (HFP; n=45) and low (LFP; n=41) feather pecking lines of laying hens. The system was well able to detect the higher activity level in the HFP line compared with the LFP line (10 vs 5 m moving distance), that was also found in previous studies. Interestingly, within the HFP line, birds phenotyped as feather peckers using traditional video observations were found to be the most active individuals compared with the other phenotypes. Precision phenotyping technology could become an important tool to automatically monitor health and welfare of individual animals housed in groups.

## The impact of Northern fowl mite infestation (*Ornithonyssus sylviarum*) on nocturnal behavior of White Leghorn laying hens

*Leonie Jacobs[1], Giuseppe Vezzoli[2], Bonne Beerda[3] and Joy Mench[4]*
*[1]Virginia Tech, Department of Animal and Poultry Sciences, 175 W Campus Drive, Blacksburg, VA 24061, USA, [2]College of the Desert, School of Mathematics and Sciences, 43-500 Monterey Avenue, Palm Desert, CA 92260, USA, [3]Wageningen University, Department of Animal Sciences, De Elst 1, 6708 WD Wageningen, the Netherlands, [4]University of California, Davis, Department of Animal Science, 1 Shields Ave, Davis, CA 95616, USA; jacobsl@vt.edu*

Northern fowl mites (*Ornithonyssus sylviarum*) are the primary ectoparasites in caged laying hens in The United States. High levels of infestation can lead to reduced production, cause irritation, anemia and even lead to mortality. Northern fowl mites potentially disrupt resting and sleep. We investigated the impact of *Ornithonyssus sylviarum* infestation on nighttime behavior of laying hens. Sixteen beak-trimmed White Leghorn hens were caged individually with visual, olfactory and auditory contact with conspecifics. Cages contained a dust box, however perches were not provided. Hens were experimentally inoculated with approximately 35 mites at 25 weeks of age and observed for "dozing" (withdrawn neck and tail down), "sleeping" (head tucked into feathers or behind wing, tail down, feathers fluffed), "preening" (raise feathers, clean and realign with beak) and "active" (any other activity whilst awake). Focal continuous observations were made from infrared video recordings taken from 22:00 h until 06:00 h (dark period) for two consecutive nights at pre-infestation week 0 (25 weeks old) and post-infestation weeks 3, 5 and 7 (respectively, 28, 31 and 33 weeks old). Hens were used as their own controls, and data were analyzed with linear mixed models, with length week of infestation as independent variable and frequency and duration of behaviors as dependent variables. Results from pre-infestation week 0 (no mites) and post-infestation week 3 (mean of 676 mites per hen) show that hens spent most time dozing (85% of time; 13 bouts per night), followed by sleeping (6.3%; 7 bouts), preening (2.4%; 10 bouts) and active (0.9%; 8 bouts). When infested (week 3), hens preened more frequently and for longer than they did prior to infestation (week 0). Preening frequency increased from 7.6 bouts/night in week 0 to 10.4 in week 3 ($P<0.01$) and preening duration increased from 68 sec/h in week 0 to 92 sec/h in week 3 ($P<0.01$). Infestation also increased dozing frequency from 10.7 (week 0) to 13.6 bouts (week 3; $P<0.01$) without affecting total dozing duration. Other behaviors were unchanged by infestation. Our preliminary analysis thus shows that even low levels of mite infestation increased the frequency and duration of preening and shortened hens' dozing bouts, given the increased frequency and unchanged duration. These observations may indicate sleep disruption due to irritation.

**Scratching that itch: dairy cows with mange work harder to access a mechanical brush**

*Ana Carolina Moncada, Heather W. Neave, Marina A.G. Von Keyserlingk and Daniel M. Weary*
*University Of British Columbia, Animal Welfare Program, Faculty of Land and Food Systems,*
*2357 Main Mall, Canada V6T 1Z4 Vancouver BC, Canada; ana.moncada@alumni.ubc.ca*

Engaging in grooming behaviours is important for animals to maintain skin cleanliness and health by reducing pathogen loads, such as bacteria, fungi and ectoparasites. Indoor-housed cattle are frequently infested with mites that can cause a localized inflammatory response and skin damage (known as mange). Mange is also thought to cause itchiness that could potentially be relieved if cows are provided with an opportunity to groom or scratch. Unfortunately, these opportunities are often lacking for indoor-housed dairy cattle. The aim of this study was to determine if dairy cows with mange decrease their motivation to groom after an anti-parasite treatment. Cows (n=24) with signs of severe mange (scaled and crusted skin lesions on the tail head) were trained to access a mechanical brush by pushing on a weighted gate. We hypothesized that the maximum weight pushed by the cows would decline after antiparasite application. Cows were first tested before treatment for a total of 9 d (3 d at 20 kg, 3 d at 40 kg, then 3 d at 60 kg). Cows were treated with a pour-on anti-parasite (EPRINEX Merial, 500 µg/kg BW) and were then re-tested with the weighted gate as described above for a 9 d period (post-treatment period 1) and then for an additional 9 d period (post-treatment period 2) to assess the longer-term effects of the anti-parasite application. Once a cow gained access she was permitted unlimited use of the brush. Only 16 cows met the initial training criterion (successfully pushed the 20 kg weighted gate during pre-treatment) and were included in the final analysis. Total daily brushing time was calculated from video recordings, and averaged within each weight (20, 40, and 60 kg) for each treatment period (pre-treatment, post-treatment period 1, and post-treatment period 2). The effect of treatment on total daily brushing time was analyzed in a repeated measures model with cow specified as subject. Values were square-root transformed and reported as mean and confidence intervals after back-transformation. Cows spent a similar amount of time brushing even when the gate was the heaviest (mean: 8.6; CI: 5.7-12.0 min; $P=0.74$). Cows tended to reduce daily brushing time in the initial period after treatment (mean: 7.6; CI: 4.4-11.6 min; $P=0.1$), and significantly reduced daily brushing time in the second period after treatment (mean: 6.5; CI: 3.6-10.3 min; $P=0.01$) compared to before treatment (mean: 12.1; CI: 7.9-17.1 min). These results provide evidence that cows with mange are more motivated to scratch when experiencing a parasite infestation than after treatment. Thus, opportunities to groom in indoor housing environments may be especially important for dairy cows with mange.

## Automated feeding behaviors associated with subclinical lung lesions in pre-weaned dairy calves

*M. Caitlin Cramer[1], Theresa L. Ollivett[2] and Kathryn L. Proudfoot[3]*
[1]*University of Wisconsin-Madison, 1675 Observatory Dr., Madison, WI 53706, USA,* [2]*University of Wisconsin-Madison, School of Veterinary Medicine, 1015 Linden Dr., Madison, WI 53706, USA,* [3]*The Ohio State University, College of Veterinary Medicine, 1920 Coffey Rd., Columbus, OH 43210, USA; catie.cramer@gmail.com*

Calves with clinical bovine respiratory disease (BRD) have been found to have slower drinking speed and fewer unrewarded feeder visits compared to calves without BRD. However, lung lesions, identified using ultrasonography, can be present with or without clinical signs; it is unknown if subclinical lung lesions (SLL) affect calf behavior. The objective was to determine if calves with SLL exhibit differences in feeding behavior from calves with clinical or no BRD. Pre-weaned dairy calves (n=133) were enrolled upon entry (21±6 d old) to a group-housed automated milk feeder barn. Researchers performed twice weekly health exams, which included a clinical respiratory score (CRS; - or +; calculated with nasal, eye, ear, cough, and rectal temperature scores) and a lung ultrasound score (0 to 5, based on severity of lung lesion). BRD status for each calf was defined as SLL (calves with any lung lesion and CRS-; n=81) or clinical (CRS+ with or without lung lesions; n=39), based on the first BRD event. Normal calves (n=13) never had lung lesions or CRS+. Feeding behavior data were collected automatically during the 3 d before and the day of diagnosis (d 0; 4 d total) for SLL and clinical calves; for normal calves, d 0 was set using the mean age of all calves with SLL and clinical BRD (31±8 d old). Three mixed linear models (drinking speed, daily milk intake, and average meal size) and two logistic models (Poisson distribution; no. of rewarded or unrewarded visits) were used to determine if BRD status was associated with feeding behaviors. All models included day relative to diagnosis as a repeated measure, and sex, breed (Holstein or Jersey), milk replacer type (medicated or probiotic), BRD status, and day by BRD interaction as fixed effects; calf was included as a random effect. There was an effect of BRD status on milk intake ($P=0.01$), which was driven by SLL calves drinking more milk than both clinical (10.9 vs 9.9 l/d; $P=0.04$) and normal calves (10.9 vs 8.9 l/d; $P=0.04$). There was no difference in milk intake between clinical and normal calves (9.8 vs 8.9 l/d; $P=0.31$). There was an effect of BRD status on drinking speed ($P=0.01$), whereby SLL calves drank faster than clinical calves (767.9 vs 664.0 ml/min; $P=0.0004$), and clinical calves tended to drink slower than normal calves (664.0 vs 771.9 ml/ min; $P=0.07$). There was no difference in drinking speed between SLL and normal calves (767.9 vs 771.9 ml/min; $P=0.9$). There was no effect of BRD status on any other behavior. There were also no effects of day relative to diagnosis, and no interactions between day and BRD status ($P>0.05$) for any behavior. We found that calves with SLL drank faster than clinical calves, which was expected. However, SLL calves had similar drinking speeds to normal calves and drank more milk than both clinical and normal calves, which was unexpected. Further research is needed to determine if lung lesion severity affects differences in behavior before and after diagnosis.

## Changes in social behaviour in dairy calves infected with *Mannheimia haemolytica*

*Catherine L. Hixson[1], Peter D. Krawczel[2], J. Marc Caldwell[3] and Emily K. Miller-Cushon[1]*
*[1]University of Florida, Department of Animal Sciences, 2250 Shealy Drive, Gainesville, FL 32611, USA, [2]University of Tennessee, Department of Animal Science, 2506 River Drive, Knoxville, TN 37996, USA, [3]University of Tennessee, Large Animal Clinical Sciences, 2406 River Drive, Knoxville, TN 37996, USA; catlhixs@ufl.edu*

Insight into behavioural changes associated with disease may be a means to improving identification, management, and welfare of sick animals. The objective of this study was to identify effects of illness on social behavior in group-housed Holstein dairy calves. We used a previously-validated mild experimental disease challenge model with *Mannheimia haemolytica* (MH), a main cause of bovine respiratory disease. Calves (aged 3-7 weeks) were group-housed based on age (6 calves/pen, 6.6 m$^2$/calf) and provided pasteurized waste milk (8 l/d) twice daily. Within group, calves were randomly assigned to 1 of 2 treatments: (1) inoculation at the tracheal bifurcation with $3\times10^9$ cfu of MH suspended in 5 ml of phosphate buffered saline (PBS) followed by a 120 ml wash PBS (MH; n=12, 3 calves/pen); or (2) inoculation with 5 ml + 120 ml of PBS only (Control; n=12, 3 calves/pen). Rectal temperatures were collected twice daily and social behaviour, including social proximity, contacts, and allogrooming, were characterized from video during daylight hours (15 h/d) on d 0 to d +2. At 10 d post-challenge, all calves appeared clinically normal and were treated with an antimicrobial. Data were calculated as hourly totals and summarized by day. Data were analyzed in a general linear repeated measures mixed model. Rectal temperatures of calves inoculated with MH were elevated throughout the challenge compared to Control calves, peaking at 12 h post-inoculation (39.2 vs 38.9 °C; SE=0.06; P=0.015), indicating a mild disease state. Social lying time (<1 body length) did not differ between treatments (30.5 min/h; SE=1.6; P=0.59) but social lying bout frequency was subject to a treatment by day interaction (P=0.004), occurring less frequently for MH calves on d 0 (0.44 vs 0.75 bouts/h; SE=0.07; P=0.001) compared to healthy penmates. Calves inoculated with MH tended to initiate social grooming less frequently (3.4 vs 4.7 bouts/h; SE=0.49; P=0.07) but received longer periods of grooming from penmates (0.43 vs 0.35 min/bout; SE=0.028; P=0.042) over the observation period. All calves received a similar duration of social contact from penmates (1.3 min/h; SE=0.20; P=0.94) but MH calves received less frequent (9.68 vs 11.2 contacts/h; SE=0.56; P=0.017) but longer periods of contact (0.16 vs 0.10 min/contact; SE=0.024; P=0.011), and initiated fewer contacts (8.61 vs 13.6 contacts/h; SE=0.89; P<0.001). In conclusion, calves exposed to this disease challenge model altered their social behaviour, reducing initiation of social interactions, and received longer periods of grooming and contact from penmates. Changes in social interactions may be useful indicators of early stages of disease in group-housed calves.

## Can calving assistance influence newborn dairy calf lying time?

*Marianne Villettaz Robichaud[1,2], David Pearl[1], Jeffrey Rushen[3], Sandra Godden[4], Stephen Leblanc[1], Anne Marie De Passillé[3] and Derek Haley[1]*
*[1]University of Guelph, Ontario Veterinary College, Population Medicine, 50 Stone Road E., Guelph, N1G 2W1, Canada, [2]Université Laval, Sciences animales, 2425 rue de l'Agriculture, Québec, G1V0A6, Canada, [3]University of British Columbia, 2357 Main Mall, Vancouver, V6T 1Z4, Canada, [4]University of Minnesota, College of veterinary medicine, Veterinary population medicine, 1365 Gortner Avenue, St. Paul, MN 55108, USA; marianne.villettaz@gmail.com*

To minimize morbidity and mortality associated with calving, dairy producers and veterinarians often provide assistance to their cows and heifers. Calving is believed to be painful for cows and calves, especially when assistance is provided. In behavioral research, lying behavior is sometimes measured as a non-invasive indicator of discomfort and pain in dairy cattle. The aim of our study was to evaluate the impacts of calving characteristics on the lying behavior of newborn dairy calves. As part of a randomized controlled trial on the timing of calving assistance, the lying behavior of 85 newborn female Holstein dairy calves was measured continuously for 10 days after birth using accelerometers. Of the 85 calves, 47 were born unassisted and 38 received assistance during calving. Within the 38 assisted calves, 29 received early assistance, defined as pulling the calf approximately 15 min after first appearance of both hooves, and 9 received late assistance, defined as pulling the calf 60 min after first appearance of both hooves. For 37 assisted calves, information on the duration of the pull and the average force applied during the assistance was recorded using an electronic load cell. The associations between calving characteristics and lying time were analyzed using univariable and multivariable mixed models that included a random intercept for animal and random slope for days post-calving. We hypothesized that calves receiving greater levels of assistance would have longer lying times in the days following birth, suggesting greater discomfort. Over the first 10 days of life, the relationship between days since birth and lying time was quadratic with lying time declining sharply over the first 7 days ($P<0.001$). Total lying time was not significantly different between the calves that received assistance compared to those born unassisted ($P>0.05$). Among the 38 assisted calves, the timing of the assistance (early vs late) did not significantly influence the calves daily lying time ($P>0.05$). However, when examining the effects of the length and force of pulling, we found there was a significant interaction between these two factors on calf lying time ($P=0.034$); the lying time of calves pulled with less than 40 kg of force increased as the length of pulling increased, but decreased for calves pulled with 40 kg of force or more. These results suggest that in dairy calves the duration and strength of pulling during assistance have more effect on lying behavior (and possibly levels of discomfort/pain) than the timing of assistance.

**Fitness for transport of cull dairy cows at auction markets in British Columbia**

*Jane Stojkov and David Fraser*
*The University of British Columbia, Faculty of Land and Food Systems, Animal Welfare Program,*
*2357 Main Mall, V6T 1Z4 Vancouver, BC, Canada; stojkov@mail.ubc.ca*

In North America, dairy cows are regularly removed from dairy herds and enter the marketing system which involves transportation to public auction markets where the cows are marketed in a sale ring and purchased by a slaughter plant buyer. Many cull dairy cows are removed from the herds because of health reasons and their fitness for transport may vary because of seasonal differences in health, delayed (or poor) culling decisions, injuries during transport and other factors. Moreover, dairy producers lack feedback on the removed cows', including their fitness for transport and how certain conditions influence the price. Therefore, the objectives of this study were (1) to evaluate the condition of cull dairy cows sold at auction markets, (2) to test the seasonal effect on the cows' fitness for transport, and (3) to quantify how fitness for transport affects the price paid. Two auction markets in British Columbia, Canada were visited regularly from May 2017 to January 2018. Trained assessors were present at 115 sales and observed 4,641 cull dairy cows while they were walked/sold in the auction ring. Assessors assessed the animals' body condition score (BCS), lameness score (LS), udder condition, and injuries, plus the cows' weight and price. Logistic regression was used to test the seasonal effect on cows' fitness for transport, and a general linear model was used to assess how the animal's condition influenced price. Approximately 8% of the cows were thin (BCS≤2), 8% were clinically or severely lame (lameness score ≥4), 13% had engorged and/or inflamed udder, and 8% had other fitness-related injuries including abscesses (2.2% of cows), hobbles (0.7%), signs of pneumonia (0.3%), eye injury (0.3%), and lump jaw (0.1%). Season had an overall significant effect on the fitness for transport ($P<0.0001$). Cows removed from the dairy herds were more likely to be unfit if removed in the fall (OR 1.54, 95% CI 1.30-1.83, $P<0.0001$) and summer (OR 1.20, 95% CI 1.01-1.44, $P<0.04$). The price was most reduced if cows were thin (BCS≤2) or had fitness-related injuries (-47.7±2.1 and -47.3±2.0 ¢/kg, respectively; $P<0.0001$). Prices were reduced to a lesser degree by lameness score ≥4 (-29.4±2.0 ¢/kg, $P<0.0001$), and by udder condition (-15.1±1.6 ¢/kg, $P<0.0001$). Body weight had a positive effect on price: each kg increase in body weight increased the price by 0.05±0.003 ¢/kg ($P<0.0001$). This study will provide information to producers about the fitness for transport of the cows shipped to auction markets, the seasonal effect of certain conditions, and information about the effect of fitness for transport on price. This information could help producers make better culling decisions and may encourage them to consider alternative options such as sending cows directly to the abattoir, emergency on-farm slaughter, or on-farm euthanasia.

## Plasma cortisol concentration, immune status and health in newborn Girolando calves

*Paula P. Valente[1], João A. Negrão[2], Euclides B. Malheiros[1], José J. Fagliari[1] and Mateus J.R. Paranhos Da Costa[1]*
[1]*Faculty of Agricultural and Veterinary Sciences, UNESP, 14.884-900, Jaboticabal, SP, Brazil,*
[2]*Faculty of Animal Science and Food Engineering, USP, 13635-900 Pirassununga, SP, Brazil;*
*paula13v@hotmail.com*

The aim of this study was to evaluate the relationship between plasma cortisol concentrations during the first days of life of Girolando calves and their immune status and health. Twenty-four Girolando calves from a normal parturition were studied, each receiving a protocol to ensure appropriate navel care and ingestion of at least 4 l of good quality colostrum (with densities 1,045 to 1,075 $g/cm^3$) within the first 6 hours of life. Calves were separated from their mother approximately 12 hours after birth, and moved to in an individual outdoor housing system, where they were offered 6 l of milk in 2 daily feedings until 35 days old. Concentrate was freely available from the 2nd day of life. Blood samples were collected in the first and third days the calves' life, always in the morning. The serum concentrations of immunoglobulins IgG and IgA, ceruloplasmin, haptoglobin, $α_1$-acid glycoprotein and transferrin were measured by electrophoresis by use of SDS-PAGE. Cortisol was measured by Elisa, while haemoglobin concentration, total counts of erythrocytes and leukocytes were assessed with a semi-automatic counter and blood smears were done to calculate the neutrophil:lymphocyte ratio (N:L). The plasma concentration of cortisol varied among the calves, but for all of them it was within the range for calves delivered by normal parturition in other studies, and lower than those reported for calves from dystocic parturition. Quartiles were used to define four classes (C) of calves according to the calves' plasma cortisol concentration (1st and 3rd day), as follows: C1) ≤41.16 ng/ml; C2) >41.16 and ≤52.46 ng/ml; C3) >52.46 and ≤93.49 ng/ml; and C4) >93.49 ng/ml. Analyses of variance were performed using PROC MIXED of SAS software package; the model included the fixed effects of class, sampling time and the interaction between class and sampling time on dependent variables. Significant effects of class ($P<0.05$) were found for the concentrations of IgG, total plasma protein, haptoglobin, and for N:L. C4 showed the highest means of IgG and total protein concentrations (3,244.58±202.31 mg/dl and 8.37±0.30 g/dl, respectively) while C1 had the lowest ones (2,275.1±177.69 mg/dl and 6.98±0.29 g/dl). This result implies that animals from C4 had better immune status and were better prepared to face the environmental challenges. Moreover, C2 calves showed the highest ratio for N:L and haptoglobin concentration (2.60±0.24 and 13.28±1.45 mg/dl, respectively), and C4 the lowest (1.70±0.27 and 7.61±1.44 mg/dl, respectively); haptoglobin concentrations of C1 (11.17±1.44 mg/dl) and C2 did not differ ($P>0.05$). These results indicate that animals from C1 and C2 could be suffering an inflammatory process, and consequently facing a welfare problem. Based on these results, we conclude that plasma cortisol concentration might have an important role in the success of the newborn calves' adaptation to the environmental challenges during their first days of life. Financial support was given by FAPESP (Process 20015/00606-1).

**Behavioural and physiological responses as early indicators of disease in New Zealand dairy calves**

*Gemma L. Lowe[1], Mairi Stewart[2], Mhairi A. Sutherland[3] and Joseph R. Waas[1]*
*[1]University of Waikato, School of Science, The University of Waikato, Private Bag 3105, Hamilton 3240, New Zealand, [2]Greyhound Racing New Zealand, P.O. Box 38313, Wellington Mail Centre, Lower Hutt 5045, New Zealand, [3]AgResearch, Animal Behaviour & Welfare, 10 Bisley Rd, Ruakura, 3216 Hamilton, New Zealand; gll1@students.waikato.ac.nz*

Automated systems are needed to monitor animal health and welfare, and detect disease before overt clinical signs are evident. This study was part of a larger project investigating automated methods with a focus on the use of infrared thermography for early disease detection of neonatal calf diarrhea (NCD). This part of the study investigated physiological and behavioural responses associated with NCD onset in calves experimentally infected with rotavirus; and assessed the suitability of these responses as indicators for early disease detection. Forty-three calves were either: (1) infected with rotavirus (through an oral drench (40 ml water and 6 ml faeces (from rotavirus positive calves)) at 6 days of age (n=20), or (2) acted as uninfected controls (n=23). Control and infected animals were housed separately, but handled in the same manner. Daily assessments of coat condition, gut fill, faecal consistency, rectal temperature and dehydration levels were conducted. Once exhibiting clinical signs (e.g. scouring and dehydration), faecal samples were collected to verify NCD as the cause of illness and calves were treated with antibiotics and electrolytes. Lying behaviour was recorded continuously using accelerometers. Respiration rate (RR) was recorded daily by observing flank movements. Drinking behaviour at the water trough was filmed continuously to determine the number and duration of visits. An outbreak of *Salmonella* (NCD causing pathogen) meant all calves developed NCD; therefore treatment was ignored and each animal was analysed as its own control with data analysed relevant to when each calf displayed clinical signs of NCD regardless of the causative pathogen. A sign test measured the significance of changes between periods (days -7 to -4 to days -3 to 0 and days -7 to -1 (pre) to days 0 to 6 (post) relative to clinical signs (day 0)) and the standard error of the difference (SED) measured variability. There was no change in RR or lying time prior to clinical signs of disease, but both decreased following clinical signs of disease (34.9±7.3 vs 29.9±9.1 breaths/min, 1,085.3±61.2 vs 1,041.5±91.0 min/day respectively: $P<0.001$). Number of lying bouts decreased (16±2.9 vs 14.7±3.3 bouts/day: $P=0.017$) and bout duration increased (71.0±16.1 vs 79.1±19.0 min/bout: $P<0.001$) prior to and following clinical signs of disease (15.5±2.9 vs 13.4±3.1 bouts/day, 74.5±15.5 vs 85.3±21.2 min/bout respectively: $P<0.001$). There was no change in number of visits to the water trough, but visit duration increased prior to clinical signs of disease (22.0±16.0 vs 27.0±14.0 sec/visit: $P=0.027$). In conclusion, number and duration of lying bouts, and duration of water trough visits, show potential as early indicators of disease. Integrating these measures into an automated system has the potential to alert farmers to disease onset, enabling earlier treatment and isolation of diseased animals. Such technology has the potential to reduce production costs and improve calf welfare.

## Postpartum social behaviour – differences between sick and healthy Holstein cows

*Borbala Foris[1,2], Julia Lomb[2], Nina Melzer[1], Daniel M. Weary[2] and Marina A.G. Von Keyserlingk[2]*
*[1]Leibniz Institute for Farm Animal Biology (FBN), Institute of Genetics and Biometry, Wilhelm-Stahl-Allee 2, 18196 Dummerstorf, Germany, [2]University of British Columbia, Faculty of Land and Food Systems, Animal Welfare Program, 2357 Main Mall, Vancouver, BC, V6T 1Z4, Canada; foris@fbn-dummerstorf.de*

The weeks after calving represent a highly sensitive period for dairy cows characterized by metabolic changes, negative energy balance and high disease prevalence. Previous work suggests that recognizing behavioural changes related to illness may help to identify cows with a higher risk of developing clinical signs. When animals become ill, they may be less willing to engage in agonistic and affiliative social interactions. Our aim was to compare sick and healthy cows in agonistic displacements and affiliative grooming. We followed a dynamic group of 20 Holstein cows (mixed parity, most cows within 3 weeks after calving) via continuous video observation for two 2-day periods (P1, P2), 10 days apart. During the first 21 days after calving, rectal temperature was taken daily and cows were examined for metritis on every third day. Other clinical diseases (milk fever, ketosis, mastitis) were diagnosed and recorded following farm protocol. In this study a cow was considered sick from the day before diagnosis of until the end of treatment and clinical signs. During P1 five cows were sick (metritis or fever) and 15 cows were healthy. During P2 the five previously sick cows were healthy and 10 of the original (healthy) group members were still present in the group. Social interactions were continuously recorded for all cows in the whole pen. Frequencies of social interactions were used for further analysis. Based on displacements we calculated a dominance score for each cow via the normalized David's score (DS). We also calculated an index for grooming (GI; the ratio of received grooming to all grooming a cow was involved in). Sick and healthy cows were compared at both time points using the Wilcoxon rank-sum test. Spearman's rank correlation between the two periods was used to assess the stability of DS and GI over time. We observed 65 grooming and 323 displacements during P1 and 84 grooming and 337 displacements during P2. During P1, sick cows had a lower GI than healthy cows ($P=0.042$), but they did not differ in DS. During P2, neither DS nor GI differed between previously sick and healthy cows. A cow's DS during P1 was strongly associated with her DS during P2 ($R_s=0.93$). This association was moderate for GI ($R_s=0.57$), but higher if only the 10 healthy cows were considered ($R_s=0.77$). These results indicate that sick cows are less often recipients of grooming during illness, and that this difference disappears once animals recover. We speculate that healthy cows may avoid grooming sick ones to reduce the risk of disease transmission, or that sick cows seek social isolation and therefore receive less grooming. Changes in the grooming index between periods for sick cows, and the stability of grooming index for healthy cows, suggest that grooming behaviour may be a useful indicator of illness.

**Cow behaviour and productivity prior to health disorder diagnoses in dairy cows in robotic milking herds**

*Meagan King[1], Stephen Leblanc[2], Tom Wright[1], Ed Pajor[3] and Trevor Devries[1]*
*[1]University of Guelph, Department of Animal Biosciences, 50 Stone Road E, Guelph, ON, N1G 2W1, Canada, [2]University of Guelph, Department of Population Medicine, 50 Stone Road E, Guelph, ON, N1G 2W1, Canada, [3]University of Calgary, Department of Production Animal Health, Faculty of Veterinary Medicine, 2500 University Dr NW, Calgary, AB, T2N 1N4, Canada; mking08@uoguelph.ca*

We examined associations of electronically-recorded production and behaviour data before diagnosis of health disorders in cows in robotic milking herds. For 605 cows in 9 commercial herds, milking systems collected milk data while electronic collars collected rumination time and activity data. Accounting for parity and days in milk (DIM), we examined data relative to the day of diagnosis for health disorders occurring in the absence of, or at least 14 d before, another disorder: mastitis (n=13), new cases of lameness (n=45), and subclinical ketosis (SCK; n=113). However, all cases of displaced abomasum (DA; n=8) occurred in conjunction with other disorders. For each illness separately, day relative to diagnosis and health status were examined as fixed effects in generalized mixed linear regression models. Using a Tukey's test procedure, we determined each variable's baseline trajectory and deviations that sick cows exhibited from it, as well as differences from a group of healthy cows (no health disorders in first 50 DIM) and an average group of all cows, who were given mock diagnosis days using the mean DIM at diagnosis for each disorder. On 6 to 14 d of the 2 wk before diagnosis, cows with DA or mastitis had lower milk yield, rumination time, milking frequency, activity, and milk temperature than healthy cows, as well as deviations from their own baseline data. For cows with DA, milk production and rumination time deviated from baseline trajectory 12 d before diagnosis, declining by 1.3 kg/d ($P<0.001$) and 19 min/d ($P<0.001$), respectively. Those cows also had deviations in activity and milk temperature. Cows with mastitis had greater maximum milk conductivity than healthy cows and deviated 12 d before diagnosis by +1.4 mS/d ($P<0.001$). Rumination time and milk yield of mastitic cows deviated from their baseline trajectory at 8 and 7 d before diagnosis, respectively, dropping by 10 min/d ($P<0.001$) and 1.5 kg/d ($P<0.001$). Compared to healthy cows, those with SCK or new cases of lameness generally had lower milk yield, rumination time, milk temperature, supplement intake, and milking and refusal frequencies. Only the milk temperature of lame cows deviated from baseline. Therefore, acute health disorders (DA and mastitis) were associated with sharp deviations from baseline behaviour and production data, whereas more chronic disorders (SCK and lameness) were associated with significant, but subtle, longer-term changes. Because sick cows differed from the healthy group before they deviated from their own baseline and the average of all other cows, including a healthy reference group in health alerts could refine the ability of detection models to identify subtle deviations in early lactation. The variability between cows' responses to illness merits further investigation into how to best use these variables for automated illness detection.

**Feeding preferences for tannin-rich forage in Holstein cattle artificially infected with gastrointestinal nematodes**

*Aubrie Willmott-Johnson[1], Simon Lachance[1], Andrew Peregrine[1], John Gilleard[2], Elizabeth Redman[2], James Muir[3] and Renée Bergeron[1]*
[1]*University of Guelph, Guelph, Ontario, N1G 2W1, Canada,* [2]*University of Calgary, Calgary, Alberta, T2N 1n4, Canada,* [3]*Texas A&M University, Stephenville, Texas, 76401, USA; rbergero@uoguelph.ca*

Gastrointestinal nematode (GIN) infections in young dairy cattle can lead to lower growth rates, decreased productivity, and discomfort. Following the rise of anthelmintic drug resistance, condensed tannin (CT)-rich forages as a natural alternative treatment for GIN infections in ruminants are being investigated. Evidence of self-medication exists in small ruminants infected with GIN; the objective of the present study was to test whether young dairy cattle infected with GIN show a preference for birdsfoot trefoil, rich in secondary compounds such as CT. Thirty Holstein heifers and steers blocked by weight and housed in 2 separate pens (average weight ± SD: pen 1 = 319.5±81.8 kg, pen 2 = 154.6±26.1 kg) were tested for preference between three forage options (A = alfalfa hay, B = birdsfoot trefoil hay, and G = grass hay) before and after artificial infection with GIN larvae. Each preference test was conducted over 5 days and consisted of 2 days of adaptation to experimental forages followed by 3 days of two-choice preference tests. During adaptation (no-choice phase), animals were individually offered each forage type as a single option, ad libitum for 45 min; the quantity consumed was recorded. Over the following 3 days, each animal was individually offered a pair of forages simultaneously; the relative consumption of each forage type was used as an indication of preference. The order of forage presentation was randomly determined. Oral infection with approximately 40,000 third stage GIN larvae was done after the first set of preference tests; the evolution of infection was monitored through weekly fecal egg counts. No animal had to be removed from the study because of clinical signs of infection. Preference tests were repeated 4 weeks following infection. A mixed model with a repeated statement was used to compare ingestion of each forage type during the adaptation period. Paired T-tests were performed to compare forage ingestion within each pair during preference tests. During adaptation to each forage type, voluntary consumption was greater ($P<0.05$) for A and G than for B, both before (A=0.79, G=0.74, B=0.34, SE=0.11 kg) and after (A=1.18, G=0.94, B=0.68, SE=0.26 kg) infection, and did not differ between A and G at either time. Before infection, A was the most preferred ($P<0.05$) of the forages, whether it was offered against G (A=0.53, G=0.22, SE=0.18 kg) or B (A=0.58, B=0.11, SE=0.16 kg). The animals did not show a preference for B or G when they were offered as a pair (B=0.28, G=0.30, SE=0.12 kg). After infection, A was still the most preferred ($P<0.05$) when paired with G (A=0.72, G=0.41, SE=0.19 kg) or B (A=0.68 B=0.12, SE=0.11 kg). However, there was a shift ($P<0.05$) in preference towards B when paired with G after infection (B=0.50, G=0.30, SE=0.12 kg). These results suggest that during a GIN infection, cattle may attempt to self-medicate with CT by modifying their forage preference.

**A qualitative study investigating opinions of pig producers; antimicrobial resistance, husbandry practices and antibiotic use**

*Kate Mala Ellen, Lauren Blake and Barbara Haesler*
*The Royal Veterinary College, The Royal Veterinary College Royal College Street, NW1 0TU London, United Kingdom; kellen4@rvc.ac.uk*

Antibiotic resistance is a type of antimicrobial resistance (AMR) that occurs when bacteria are able to survive antibiotic treatment. Overuse and misuse of antibiotics have accelerated resistance. Reduction of antibiotic use in human and animal medicine is encouraged, to safeguard the effectiveness of current drugs and ease the pressure of developing new ones. Use in the UK pig farming industry has decreased but routine use is still common practice. Three farms were visited (two conventional and one organic); semi-structured interviews and observations were conducted with the farmers to find out their perceptions and understanding of AMR, how antibiotics are used, what drives use, opinions of different husbandry practices and the factors affecting antibiotic reduction. Voice recordings were transcribed verbatim by the author. Transcripts were annotated and thematic analysis performed through coding. Thematic findings were compared, categorised and revised in accordance with the research questions. Quotes from the empirical data support the common themes identified as influences on antibiotic use: advice of veterinarians, "we base our choice of drugs on the [vets'] recommendation"; withdrawal periods of antibiotics; strength/size of piglets at weaning, "if you wean a small, weaker pig they are more prone to bugs"; pig breed; teeth grinding and tail docking procedures, "could probably get away without amoxicillin if you weren't teething and tailing"; vaccine use, "we do use vaccines…if we didn't, we'd be using more antibiotics"; building/ground sterility and maintenance; assurance scheme standards "people are so confused…if they were asked what the Red Tractor was, I'd be surprised if more than 25% really knew what it meant"; and consumer choice, "help drive the British market". The findings gave insight into farmers' opinions and highlighted factors to address in order to change antibiotic use behaviour. They have provided much to be learnt from and to be kept in mind when reducing antibiotic use in the UK.

## Assessing positive welfare in broiler chickens: effects of specific environmental enrichments and environmental complexity

*Judit Vas[1], Neila Ben Sassi[2], Guro Vasdal[3] and Ruth C. Newberry[1]*
*[1]Norwegian University of Life Sciences, Faculty of Biosciences, Department of Animal and Aquacultural Sciences, P.O. Box 5003, 1432 Ås, Norway, [2]Neiker-Tecnalia, Campus Agroalimentario de Arkaute, Apto 46, 01080 Vitoria-Gasteiz, Spain, [3]Norwegian Meat and Poultry Research Centre, Lørenveien 38, P.O. Box 396 Økern, 0513 Oslo, Norway; judit_banfine.vas@nmbu.no*

Many studies comparing poultry welfare in different housing systems report about differences in prevalence of problems whereas field studies on positive effects of environmental enrichments in commercial broiler flocks are scarce. The aim of this study was to examine the impact of specific enrichments and overall environmental complexity on potential indicators of positive affective states in broiler chickens close to slaughter age. We focused on elements of exploratory, play and comfort behaviour as evidence of positive (reward-based) as opposed to negative (aversive) welfare states. Thirty Norwegian commercial broiler flocks varying in number and types of enrichments (woodshavings bales, peat, boxes) were visited when the birds (mixed sex Ross 308) were 28±1 days old. Eight flocks received no enrichments, 16 flocks received half-bales of compressed woodshavings (for foraging and sitting on top and around), 19 received peat (for foraging and dustbathing), and 22 were provided with boxes (plastic crates or cardboard boxes, for sitting on top, around and underneath). Environmental complexity was calculated as the number of these environmental enrichment types provided (range 0-3). With a newly developed "behaviour transect" method, we made 15-s scans of birds within approximately 96 observation patches per house (observing an estimated 4,745±932 birds per flock, in 317±55 $m^2$ of the house area). We recorded the number of birds per patch per 15-s scan engaged in five mutually-exclusive behaviours: ground-scratching (an element of exploratory behaviour), play-fighting, running, and wing-flapping (play-related behaviours) and vertical wing shakes (the most visible element of dustbathing, representing comfort behaviour). Generalized linear mixed models were used to evaluate effects of (1) enrichment type and (2) environmental complexity, on occurrence of these behaviours in flocks below (13.8±0.5 birds/$m^2$) and above (16.3±0.7 birds/$m^2$) the median stocking density on the day of the visit. More birds were observed running in flocks with than without woodshavings bales (0.18±0.05 vs 0.12±0.02% of birds, $P<0.001$) and in flocks with lower compared to higher stocking density (0.22±0.05 vs 0.09±0.02%, $P<0.001$). More dustbathing behaviour was observed in flocks with than without each of the specific enrichments (woodshavings bales: 0.22±0.06 vs 0.13±0.04% of birds, $P=0.048$; peat: 0.22±0.05 vs 0.09±0.03%, $P<0.001$; boxes: 0.21±0.05 vs 0.09±0.04%, $P<0.001$). Ground-scratching, play-fighting and wing-flapping were not influenced by specific enrichments or density per se. However, higher environmental complexity was associated with higher levels of play-fighting ($P=0.029$), running ($P=0.016$) and dustbathing ($P=0.004$). We conclude that both enrichment type and environmental complexity contribute to the expression of behaviours proposed to reflect positive affective states in chickens.

## Does increased environmental complexity improve leg health and welfare of broilers?

*Fernanda Tahamtani[1], Ida Pedersen[2] and Anja Riber[1]*
*[1]Aarhus University, Department of Animal Science, 8830 Tjele, Denmark, [2]University of Copenhagen, Department of Veterinary and Animal Sciences, Grønnegårdsvej 15, 1870 Frederiksberg C, Denmark; anja.riber@anis.au.dk*

Increasing environmental complexity, i.e. by providing environmental enrichment, has been suggested as a way to increase activity levels and improve leg health in fast-growing broiler chickens. The aim of this study was to investigate the effects of several types of environmental complexity on leg health and welfare of fast-growing broilers housed according to conventional Danish guidelines. Sixty pens with approximately 500 broilers in each (Ross 308), corresponding to a maximum stocking density of 40 kg/m$^2$, were used. Food and water were provided *ad libitum* with a distance of 1.5 m in between. Five environmental enrichment treatments (maize roughage, vertical panels, straw bales, elevated perforated plastic platforms at 5 cm or 30 cm height), three treatments where the standard resources had been manipulated (increased distances between feed and water (7 m and 3.5 m), 34 kg/m$^2$ maximum stocking density) and one control group were randomly assigned to each pen. The enrichments were provided from placement of the day-old chicks until slaughter, and were placed centrally in the pens. The study was performed in six blocks, each with 10 pens. In each block, one treatment was applied to two pens while the others were each applied to one pen. Two out of the 60 pens had to be excluded from the study due to flooding. At 35 days of age, 60 birds from each pen were assessed for gait score, footpad dermatitis, hock burns, plumage cleanliness, presence of scratches and presence of leg deformities (varus/valgus). There was an effect of treatment on gait score ($F_{8, 3418}$=2.20; $P$=0.02). However, the significance of the pairwise testing was lost when correcting for the large number of comparisons. Numerically, birds housed with access to a 30 cm elevated platform had the lowest frequency of gait score 0, while birds kept at a lower stocking density had the highest frequency of gait score 0. There was a significant effect of treatment on footpad dermatitis ($F_{8, 3420}$=3.93; $P$=0.0001), with birds housed with access to a 30 cm elevated platform having better scores compared to birds housed with access to straw bales ($P$=0.0001) and with increased distance between feed and water ($P$=0.011). Furthermore, birds housed with straw bales had higher footpad scores compared to birds with access to a 5 cm elevated platform ($P$=0.002). There were no observed treatment effects on scratches, plumage cleanliness, leg deformities and body weight ($P$>0.05). The observed effects of the different types of environmental enrichment and environmental modification suggest that the provision of straw, time in contact with the litter, and reduced scratching of the litter due to more time spent on locomotion may be risk factors for the development of footpad dermatitis.

## Individual distress calls as a flock-level welfare indicator

*Katherine Herborn[1], Benjamin Wilson[1], Malcolm Mitchell[2], Alan McElligott[3] and Lucy Asher[1]*
*[1]Newcastle University, Institute of Neuroscience, Newcastle Upon Tyne, NE2 4HH, United Kingdom,*
*[2]SRUC, Animal & Veterinary Sciences, Easter Bush, EH25 9RG, United Kingdom, [3]University of*
*Roehampton, Life Sciences, London, SW15 4JD, United Kingdom; lucy.asher@ncl.ac.uk*

Acoustic monitoring offers insight into animal welfare by capturing vocalizations associated with different emotional states. However, simply counting the vocalizing individuals may not accurately estimate group welfare status, as here, we demonstrate social contagion in emotions. Chicks (*Gallus gallus domesticus*) emit a repetitive, high energy 'distress' call when acutely stressed. Pharmacological studies have shown that chicks in an anxiety-like state distress call continuously, while chicks in a depression-like state call intermittently, at half the total rate. Using recordings of 20 chicks, we generated artificial stimuli that mimicked these natural call distributions. As a control, we generated a stimulus using 'contact calls' from the same individuals in an unstressed state. In the first trial, 4 new flocks of 20 chicks, housed in separate, sound-proof rooms, were exposed to each stimulus in a random order for one complete day per stimulus (15 min playback/h). On the day after each stimulus, a subset per flock were temporarily moved to a test arena to explore carry-over impacts on cognitive biases in a radial maze, where individuals were challenged to leave group safety to forage in 5 equally spaced locations. Compared to controls, chicks exposed to the anxiety-like stimulus foraged for less time in a known rewarding location, suggesting context-specific pessimism or anxiety (general linear mixed model, GLMM, n=32 chicks, ID=random, location categorical, difference to control: -10±4.8 s, $t$=1.16, $P$=0.046). However, exposure to the depression-like stimulus caused a context-general shift in behaviour, with chicks discriminating less across the gradient of potentially rewarding to unrewarding locations (GLMM, n=32, ID=random, location continuous, difference to control slope: 3.59±1.54, t=2.33, $P$=0.022). In the second trial, 3 flocks of 12 chicks were retained and filmed with a thermal camera during the 1st and 7th hourly exposure to a further day's playback per stimulus. A short-term drop in comb temperature at 1st exposure to the anxiety-like stimulus was consistent with an acute stress response (GLMM, categories <15 min before and <3, 3-15 and 30-45 min after onset, n=982 measurements, flock=random, difference to control at <3 min: -0.72±0.36 °C, t=-1.98, $P$=0.048). With recurrent exposure, however, baseline comb temperature was elevated by 1 °C by the depression-like stimulus, suggesting longer-term physiological impacts (GLMM, n=935, difference to control before onset: 1.07±0.22 °C, t=4.87, $P$<0.0001). Exposure to one chick distress calling thus caused behavioural and physiological changes consistent with a negative emotional state; and specifically anxiety vs depression. Beyond signalling individual welfare, distress calls may themselves be a flock-level welfare concern worth monitoring. Whilst acoustic analysis on farms presents several challenges, we propose an automated approach to distress call monitoring.

## The effect of calcium propionate in feeding strategies for broiler breeders on palatability and conditioned place preference

*Aitor Arrazola[1,2], Tina Widowski[1,2], Michele Guerin[1,3] and Stephanie Torrey[1,2]*
[1]*University of Guelph, Campbell Centre for the Study of Animal, Guelph, ON, Canada,* [2]*University of Guelph, Animal Biosciences, Guelph, ON, Canada,* [3]*University of Guelph, Population Medicine, Guelph, ON, Canada; aarrazol@uoguelph.ca*

Broiler breeder pullets are feed-restricted throughout rearing to avoid health-related problems attributed to fast body weight gain and to achieve a profitable laying performance. This chronic feed restriction is a significant welfare concern, and the development of alternative feeding strategies has focused on decreasing voluntary feed intake by the inclusion of appetite suppressants, such as calcium propionate (CaP). Dietary inclusion of CaP decreases feeding rate, and it has been suggested that the decreased feeding motivation is due to lower palatability, gastrointestinal discomfort, or both. The objective of this experiment was to examine the effect of CaP at the inclusion rate of alternative diets for broiler breeders on palatability and the experience of a negative subjective state. Three subsamples of Ross 308 broiler breeder pullets were used: two for the palatability test (108 pullets at 2 pullets/cage on week 3, and 24 pullets in individual cages at week 8) and one for the condition place preference test (24 pullets at 4 pullets/pen). Palatability was assessed by measuring feeding rate in 10 min with birds naïve to CaP, at weeks 3 (at 0% [control] or 1.44% CaP inclusion) and 8 (at 0% [control] or 3.19% CaP). Data were analysed by age using linear mixed regression models, with cage nested in the models and body weight as a covariate. The experience of negative subjective state was examined using a conditioned place preference test. On training days, pullets consumed two pills (160 mg of CaP/pill) followed by 20 g feed allotment. Training lasted for 90 min/pullet/day during 8 consecutive days at weeks 7 and 9, and pullets' choice was tested in a T-maze twice on two consecutive days at weeks 6, 8 and 10. Data were analysed using a linear mixed regression model, with pen nested in the model and age as a repeated measure. Pullets' preference was assessed using the binomial distribution. For the palatability assessment, pullets fed CaP at 1.44% (30.76±1.36 g) ate less in 10 min at week 3 compared to pullets fed the control diet (35.94±1.36 g; $F_{1,45}$=14.49, $P<0.001$). However, there was no significant effect of 3.19% CaP at week 8. For the negative subjective state assessment, the choice for place conditioned with the consumption of the CaP pill linearly decreased over time ($F_{1,18}$=38.27, $P<0.001$) after repetitive conditioning place training. No pullets preferred the CaP pill while the preference for the placebo increased from week 6 to weeks 8 (+17.4±5.8% of pullets [6/24]; $t_{18}$=2.91, $P=0.01$) and 10 (+30.4±8.6% of pullets [9/24]; $t_{18}$=4.53, $P<0.001$). The inclusion of 1.44% CaP reduced feeding motivation during early rearing at a low feed restriction level, and pullets were less likely to choose the place conditioned with the consumption of CaP. These results suggest that inclusion of 1.44% CaP for broiler breeder pullets can reduce palatability and induce a negative affective state.

## Hepatic damage and learning ability in broilers

*Laura Bona[1], Nienke Van Staaveren[1], Bishwo Pokharel[1], Marinus Van Krimpen[2] and Alexandra Harlander-Matauschek[1]*
[1]*University of Guelph, Department of Animal Biosciences, Ontario Agricultural College, 50 Stone Road E, Guelph, Ontario N1G 2W1, Canada,* [2]*Wageningen Livestock Research, P.O. Box 338, 6700 AH Wageningen, the Netherlands; aharland@uoguelph.ca*

Methods that reduce growth rates or lower body weights in meat birds are increasingly considered as an animal welfare friendly way of farming by society. This could possibly be achieved by lowering the protein content while increasing energy content of the diet. However, in mammals, the consumption of these low protein energy-rich (LPER) diets has been associated with increased susceptibility to metabolic diseases (e.g. hepatic damage), and adverse effects on cognition. However, the effect of LPER diets on both hepatic damage and cognition have not been investigated in domestic meat chickens. Therefore, this study investigated the effect of a LPER diet on learning ability and the occurrence of liver damage in broiler chicks. Forty female Ross broiler chicks (1 d old) were housed in floor pens and received a standard commercial diet. At 18-20 d, half of the birds were gradually introduced to a control (19% CP, 3,200 ME) or LPER (17% CP, 3,300 ME) diet which was the main diet from 21-51 d of age onward. Visual discrimination training of birds (1-10 d of age) was conducted in a Y-maze until 80% of the birds reached the learning criterion of at least 5/6 correct trials for two consecutive days. Reversal of the discrimination occurred at 38-46 d of age using the same criteria. Blood samples were collected on 17-18 d and 46 d and plasma concentration of ammonia ($NH_4$) and activity of alanine aminotransferase (ALT), aspartate aminotransferase (AST) and gamma-glutamyl transferase (GGT) were analyzed. Birds were euthanized at 52 d of age and liver hemorrhage (score 0-5) and colour (1-5) were assessed. The effect of the LPER diet on indicators of liver damage (i.e. plasma hepatic markers, liver hemorrhage and colour) and the ability to learn a visual discrimination reversal task (Y-maze) were assessed using generalized linear mixed models (PROC GLIMMIX). All chicks, regardless of diet, showed signs of liver hemorrhage, liver colour scores indicative of damage, and a high activity of AST above 230 U/l indicative of liver damage. LPER birds that successfully completed the reversal discrimination task tended to have a lower liver hemorrhagic score (0.98±0.08) and had a lower AST:ALT ratio (214.8±18.95) than birds fed a control diet (1.29±0.09, $P=0.0531$ for hemorrhagic score and 297.4±20.88, $P=0.0417$ for AST:ALT ratio, respectively). No difference was observed between LPER (6.2±0.47 sessions) and control chicks (5.5±0.49 sessions) in the number of sessions needed to complete the reversal discrimination learning task successfully, however LPER birds (3.2±0.04 kg) had lower body weights than control birds (3.5±0.04 kg, $P=0.0045$) which could have influenced their ability to complete the task (e.g. ease of locomotion, higher motivation). These results suggest that a LPER diet does not increase susceptibility to liver damage and does not impact reversal discrimination learning in broiler chicks. However, the high prevalence of liver damage in broiler chicks as shown by hemorrhage and AST activity are of concern and highlights a need for further research on liver health in broilers.

## The importance of habituation when using accelerometers to assess activity levels of turkeys

*Rachel Stevenson[1], Ji-Qin Ni[2] and Marisa Erasmus[1]*
*[1]Purdue University, Animal Science, West Lafayette, IN, 47907, USA, [2]Purdue University, Agricultural & Biological Engineering, West Lafayette, IN, 47904, USA; steve189@purdue.edu*

Animal behavior is a common indicator that farmers use to identify changes in animal health and welfare. Deviation from normal activity levels in poultry can indicate that a bird is sick, lame or injured. On large commercial farms, thousands of birds can be housed together, presenting a challenge for farmers who visually inspect birds for welfare issues. Technologies, such as accelerometers, may provide objective, non-invasive methods for detecting changes in animal behavior and welfare. However, the presence of the accelerometer itself can affect behavior and normal gait of an animal. We conducted research to validate accelerometers (AXY-3 Micro Acceleration Data Loggers) in tom turkeys with the following objectives: (1) to determine if a habituation period is necessary to prevent a response to wearing an accelerometer on the leg of tom turkeys, (2) to determine if wearing the accelerometer affects the normal gait, and (3) to evaluate age-related changes in gait. Accelerometers were attached to the leg (balanced for left and right) of turkeys with a Vetwrap bandage above the hock joint. Thirty-six male commercial turkeys were recorded walking across a Tekscan® pressure pad at 8, 10 and 12 wk. Data collected from the Tekscan® included step time, step length, step velocity, maximum force, impulse, and gait time. Birds were randomly assigned to one of five groups: accelerometer and 1 wk habituation period (AH, n=7), accelerometer and no habituation period (AN, n=5), Vetwrap (no accelerometer) and 1 wk habituation period (VH, n=8), Vetwrap (no accelerometer) and no habituation period (VN, n=10), and nothing on either leg (C, n=6). Data were analyzed using a block ANOVA model in SPSS with Tukey's test for multiple comparisons. No differences between treatment groups or age were found for step length, step velocity or gait time. Cadence tended to differ between 8 wk (93.04±10.00 steps/min) and 12 wk (75.68±15.02 steps/min; $P=0.07$). Non-habituated groups took longer to take a step each time (AN: 25.03±5.03 s; VN: 24.43±3.89 s) compared to control and habituated groups (C: 16.15±4.46 s; AH: 13.44±4.59 s; VH: 13.99±7.04 s; $P=0.04$). Maximum force was lower for non-habituated groups (AN: 84.38±2.10%; VN: 86.09±3.10%) vs control and habituated groups (C: 95.82±4.22%; AH 95.29±5.48; VH: 94.48±4.99%; $P=0.03$). Similarly, impulse was higher for non-habituated (AN: 92.44±0.73%; VN: 93.21±2.40%) vs control and habituated groups (C: 84.01±3.05; AH 87.91±1.33%; VH: 87.90±1.80%; $P=0.05$). Results revealed that birds without a habituation period differed from those that did for step time, impulse, and pressure exerted on the leg with the accelerometer when walking across a pressure pad. When using AXY-3 accelerometers with bandages, it is advisable to habituate turkeys to wearing the bandage prior to collecting data.

**Do we have effective possibilities to reduce injurious pecking in tom turkeys with intact beaks?**

*Jutta Berk, Eva Stehle and Thomas Bartels*
*Institute of Animal Welfare and Animal Husbandry, Friedrich-Loeffler-Institut, Dörnbergstr. 25/27, 29221 Celle, Germany; jutta.berk@fli.de*

Injurious pecking is an important welfare and economic issue in commercial turkey production but the development and causes of this undesirable behaviour are poorly understood. Beak trimming is presently performed to reduce the damage caused by injurious pecking in turkeys. Several factors could have an influence on the development of this damaging behaviour in turkeys. The spectral distribution and source of light as well as environmental enrichment have been shown to affect the prevalence of injuries. In Germany, beak trimming is expected to be banned in the future. Keeping turkeys with intact beaks may cause a higher prevalence of injurious pecking. The aim of the study was to investigate the impacts of light quality and barn enrichment on the prevalence of injurious pecking in tom turkeys with intact beaks. In two trials each 600 male day-old B.U.T. 6 turkeys with intact beaks were allocated to six littered floor pens (each 36 m$^2$, 2.8 toms/m$^2$). Each pen consisted of two parts, differing in their luminance intensity (20 lx vs 70 lx). The pens were illuminated by two tubular fluorescent lamps (58 W) of different color temperatures (either 3,000 K or 6,500 K). Both parts were enriched with wheat grain feeders and were connected with a passage which was equipped with two antennas for recording the frequency of changes between pens. All toms were individually banded with two leg transponders and kept for 20 weeks under otherwise common conditions. Video recordings were used to evaluate behaviour including the use of enrichment devices. Additionally, the number of toms moving from one side of the pen to the other was counted automatically. Effects of light spectrum and luminance intensity on the use of feeders were tested using GLM. The prevalence of injurious pecking in response to the use of feeders and light quality was evaluated by a generalized linear mixed Poisson model. In both trials, injurious pecking occurred more frequently in pens illuminated with warm white light (3,000 K) vs 6,500 K ($P<0.0001$). Turkeys preferred to stay in the higher illuminated pens independently of the light spectrums up to an age of about 13$^{th}$ weeks. Use of wheat grain feeders was higher at a luminance intensity of 20 lx compared to 70 lx ($P=0.0213$). The duration of use varied between 1 s and 21.4 min per tom but most frequently times up to 30 s were observed. Injurious pecking decreased with increasing use of wheat grain feeders ($P<0.0001$). The rate of exchange between both parts of the pens decreased with age. The study indicates that light quality and environmental enrichment can affect the prevalence of injurious pecking. Further studies are necessary to understand the causes for injurious pecking in fattening turkeys and to limit their effects. This study was funded by the Lower Saxony Ministry of Food, Agriculture, Consumer Protection and Regional Development.

### Learning ability in broilers: does environmental enrichment help?

*Fernanda Tahamtani[1], Ida Pedersen[2], Claire Toinon[3] and Anja Riber[1]*
*[1]Aarhus University, Department of Animal Science, Blichers Alle 20, 8830 Tjele, Denmark,*
*[2]University of Copenhagen, Department of veterinary and Animal Sciences, Grønnegårdsvej 15, 1870 Frederiksberg C, Denmark, [3]École Nationale Supérieure D'agronomie et des Industries Alimentaires, 2 Avenue de la Foret de Haye, 54505 Vandœuvre-lès-Nancy, France; fernandatahamtani@anis.au.dk*

Increasing environmental complexity, i.e. by providing environmental enrichment, has been suggested as a way to improve broiler chicken welfare. It is also known to promote the development of cognitive functions, including learning and memory. The aim of this study was to test the hypothesis that housing environments of different complexity, due to environmental enrichment or manipulation of standard resources, have an effect on learning ability in broilers. Twenty-seven pens with approximately 500 broilers in each (Ross 308), corresponding to a stocking density of 40 kg/m$^2$, were used. Commercial feed and water were provided *ad libitum*, and wood shavings were used as litter material. Each pen was randomly assigned to one of four treatments: (1) provision of straw bales, (2) provision of a 30 cm elevated platform with access ramps, (3) 34 kg/m$^2$ stocking density, (4) and a control group with no enrichment or resource manipulations. At 21 days of age, 10 randomly selected broilers from each pen were tested in a passive avoidance-learning task in experimental test pens (60×50×60 cm). The birds were kept in pairs of focal and companion bird and were deprived of feed during a 3 h habituation period. Training was performed in a series of four trials by presenting control and aversive feed on red or blue papers. The aversive feed was produced by infusing control feed with 99% Methyl Anthranilate, a non-toxic colourless liquid commonly used as bird repellent due to its bitter taste. The typical response to eating the aversive feed was bill wiping and head shaking behaviour. Following a 2 h interval, testing was performed in two trials by presenting control feed on both red and blue papers in turn. The number of pecks at the feed by the focal bird was measured per trial. The birds were expected to associate the colour of the paper on which the feed was presented with the taste of the feed and show avoidance behaviour towards the colour associated with the aversive feed but not with the control feed. There was a tendency for broilers housed with access to straw bales to have higher discrimination ratio (dr) in the passive avoidance-learning task (dr=0.65±0.2) compared to broilers in the control group (dr=0.52±0.1, $F_{3,92}$=2.22; *P*=0.091). There was no difference compared to 30 cm elevated platforms (dr=0.60±0.1) or 34 kg/m$^2$ stocking density (dr=0.58±0.2). Approximately 50% of the focal birds did not meet the criteria for inclusion, which required pecking during at least two of the training trials and the control testing trial. This resulted in a much lower sample size than expected. Therefore, it is likely that a larger sample size would have shown that increasing environmental complexity can have positive effects on promoting learning ability in broiler chickens.

## Prevalence and severity of footpad dermatitis in broiler chickens raised under different litter management in humid tropics

*Olufemi Alabi[1] and Sabainah Akinoso[2]*
[1]*Bowen University, Animal Science and Fisheries Management, P.M.B. 284, 23001 Iwo, Osun State, Nigeria,* [2]*University of Lagos, Science, Technology Education, P.M.B. 1001, University of Lagos, 21001 Lagos, Nigeria; femiatom@yahoo.com*

Deep litter housing is the commonest system adopted for rearing broiler chickens in humid tropical countries. Meanwhile, poor litter management has been linked with an increased incidence of some diseases such as coccidiosis, aspergillosis, *Eischeria coli* infection footpad dermatitis (FPD). Apart from affecting the performance characteristics of broiler chickens, FPD also affects the welfare and behavior of the chickens. However, there is a dearth of information on the extent to which litter management can be used to control FPD in broiler production in the humid tropical parts of the globe hence this work that investigated the prevalence and severity of FPD among broiler chickens under different types of litter management. Three hundred and sixty (360) Hypecco strain of broiler chickens at day old from a reputable hatchery were randomly allotted into three litter management groups (T1, T2 and T3). Each group has three replicates of 40 birds in a completely randomized design. T1 represents the control group with no packing (removal of old litter and replacing with fresh or new litter materials) and turning of litter materials, T2 is the group with weekly turning of the litter materials while T3 is the group with weekly packing and replacement of litter materials. Feed and water were given *ad libitum* while other management practices in terms of medication and vaccination were strictly observed. The experiment lasted 56 days during which data on the age at first occurrence of FPD, number of incidence at 28th day, number of incidence at 56th day, severity at 28th day, severity at 56th day, mortality rate, mobility and aggression at 56th day. Data on ages at first occurrence and number of occurrences were analyzed statistically using analysis of variance and descriptive analysis respectively while severity, mobility and aggression were rated as slightly, moderately and highly. Severity was rated according to the size of the dermatitis lesion on the footpad, mobility by the number of steps made by each bird per minute and aggression by their tendencies to peck the hand of the researcher during feeding. The results revealed that in birds on T1, FPD first occurred at 26.5±0.02 days followed by T2 (43.50±0.15 days) while the first incidence in T3 was on 52.5±0.10 days. Only birds on T1 recorded FPD at 28th day (end of starter phase) while at 56th day (end of the finisher phase), the occurrence of FPD was highest among birds on T1 (42.2%) with T3 having the lowest (6.5%). FPD was slightly severe and only occurred in T1 birds at 28th day and was highly severe at 56th day but moderately severe among T2 birds and not severe with T3 birds. Birds on T1 recorded 5.0% mortality; T2, 1.0% and none in T3 though not connected with FPD. Moreover, broilers in T1 appeared dull, inactive and slightly aggressive while T2 and T3 birds were very active and not aggressive. These results suggest that better litter management especially replacement of old litter with new can be used to control incidence and severity of FPD to improve the well-being and behavior of the broiler chickens.

### Influence of purslane extract and probiotic on gait score, hock lesions, and energy and protein utilization of broilers

*Mohammad Ghorbani and Ahmad Tatar*
*University of Khuzestan, Department of Animal Science, Ramin Agriculture and Natural Resources, Mollasani, 6341773637, Ahvaz, Iran; tatar@ramin.ac.ir*

This study evaluated the effects of purslane extract and probiotic on welfare related parameters, droppings characteristics and energy and protein utilization of broiler chickens kept at high stocking density. In a completely randomized design, 280 one-day-old broiler chicks were allocated to five treatments each with four replicates. Dietary treatments included: (1) positive control (PC; 10 chicks/m$^2$); (2) negative control (NC; 15 chicks/m$^2$); (3) NC + 500 mg/kg purslane extract (PE); (4) NC + 200 mg/kg probiotic supplementation (PS); and (5) NC+500 mg/kg PE + 200 mg/kg PS. At the end of the experiment on D 42, 10 chickens in each pen were inspected for walking ability, hock and footpad burns, and abdominal plumage condition. At 42 d. of age, excreta were collected from each pen and its pH, dry matter, volatile and nonvolatile composition were measured. To calculate energy and protein utilization (energy efficiency ratio; EER and protein efficiency ratio; PER), body weight gain and feed intake were recorded at the end of each step of the rearing period. EER and PER were calculated for each step. All data were analyzed using the GLM procedure of SAS software for analysis of variance. Differences between treatment means were tested using Duncan's multiple comparison test. Statistical significance was declared at $P \leq 0.05$. The welfare related data were obtained from an average of 10 birds in a pen, with pen as the experimental unit and analyzed by a Wilcoxon test. The results of this experiment showed that increasing the stocking density positively influenced broiler EER and PER, but negatively influenced litter moisture, gait score, foot pad dermatitis and hock burns. Also, litter and excreta moisture were significantly increased with increasing stocking density ($P<0.05$). EER and PER were increased as a result of increasing density during the starter and overall experimental periods ($P<0.05$). Birds reared at high density with feed additives had greater EER and PER than the PC group ($P<0.05$). These data indicate that use of purslane extract and probiotics in birds kept at a high stocking density did not have a clear effect on the welfare related parameters measured in this study.

**Behavioral and welfare implications of antibiotic and probiotic supplementation**

*Antonia Patt and Rachel Dennis*
*University of Maryland, Department of Animal & Avian Sciences, College Park, Maryland, 20742,*
*USA; rldennis@umd.edu*

The global modern poultry industry is facing numerous changes to feeding and supplementation regulations and restrictions. Although these differ greatly across countries and companies, the fast-paced changes in poultry feed supplements are ubiquitous, especially in regards to antibiotics, pro- and pre-biotics. Unfortunately, the majority of the research into these supplements has focused on production parameters with far less research into the welfare implications of these supplements. Still fewer studies investigate the impacts of the use of these products in young, developing birds. In the present study, 1 day old broiler chicks were given water treated with an antibiotic cocktail, probiotic cocktail, or no supplement (control). Treatment water was provided *ad libitum* for the first week of life and changed daily. Following the treatment week, all birds received untreated water for the remaining 4 weeks of the study. At week 5 of age, tonic immobility (TI) durations were measured in 1 bird per pen, and flight distances from a novel human were measured in all pens (n=12/treatment). Treatment pens were blocked by room to prevent cross-contamination (4 rooms/treatment). Body and organ weights were taken as well as neural tissue samples for monoamine analysis. Probiotic-treated birds had reduced TI durations compared with both control and antibiotic treated birds ($P<0.05$). However, flight distances were greater in antibiotic-treated birds compared with both probiotic-treated and control birds ($P<0.01$). No difference was observed in final body weight between the treatments ($P>0.1$), however, antibiotic-treated birds tended to have a greater heart weight than probiotic-treated birds ($P=0.072$) and greater spleen weight compared with control birds ($P=0.057$). Antibiotic-treated birds also expressed reduced levels of serotonin in the hypothalamic tissue compared with probiotic-treated and control birds ($P<0.05$). Sequencing analysis of the microbiome from fecal samples in this study indicates changes in the gut microflora of the three treatment populations over time. These populations were most different during early developmental stages. Our data shows the importance of investigating the effects of microbiome altering supplements on bird behavior and well-being. Additionally, we have shown that long-lasting behavioral and physiological changes can be achieved by early ingestion of antibiotic and probiotic supplements. The long-lasting impacts of early antibiotic treatment inhibit positive welfare as indicated by increased fearfulness and reduced hypothalamic serotonin.

## Brazilian catchers' perceptions about manual catching of broilers and its implications on animal welfare and carcass quality

*Victor A. De Lima[1,2], Maria Camila Ceballos[1], Carolina A. Munoz[3,4], Vanessa S. Basquerote[5], Eliana Renuncio[5] and Mateus J.R. Paranhos Da Costa[1,2]*
[1]*Grupo ETCO, FCAV-UNESP, Zootecnia, Via Paulo D. Castelane, 14884-900 Jaboticabal/SP, Brazil,* [2]*Unesp, Câmpus de Jaboticabal, Zootecnia, Via Paulo D. Castelane, 14884-900 Jaboticabal/ SP, Brazil,* [3]*Animal Welfare Science Centre, University of Melbourne, Level 2/21 Bedford St, VIC3051 North Melbourne, Australia,* [4]*Universidade Federal de Juiz de Fora, Zoologia, Rua José L. Kelmer, S/n, 36036-330 Juiz de Fora/MG, Brazil,* [5]*Veterinarian, Chapecó, SC, Brazil; lima.victor@uol.com.br*

A focus group study was conducted with a team of Brazilian broiler catchers to collect quantitative and qualitative information on their perceptions about manual catching of broilers and its implications on animal welfare and carcass quality. The objective of this study was to validate the methodology of a broader study, aiming to measure the impact of catchers' attitudes on broiler welfare and carcass quality, before and after training. The focus group ran for approximately 50 min, followed a semi-structured agenda, and comprised of nine broiler catchers, with experience in handling broilers ranging from 3 months to 10 years. The discussion offered the opportunity to identify some catchers' underlying beliefs that influence their behaviour and that, consequently, may have an impact on animal welfare and carcass quality. In general, catchers visually assess the welfare conditions of the animals before the catching process. The main indicators mentioned by the participants were: body condition, dirtiness of birds, presence of dead birds, and main aspects of birds' behaviour (vocalisation, if they are calm or agitated, and if they are spread or huddled in the sheds). Birds are carried upright one by one or in pairs and catchers acknowledged that they feel pain and that they vocalised differently when handled incorrectly. They recognised that catching the birds by the back is better for the animals and causes fewer lesions than catching birds by the legs, neck or wings. Catchers also observe the sheds' conditions before catching the birds. They mainly assess the infrastructure of the shed, the bedding condition, and the presence of any material or equipment (e.g. wood, manual feeders, feed bags and heaters), which are often perceived as barriers for the catching process. Although they are not fully aware of how farmers handle the animals, there was a strong acknowledgement that the attitudes and behaviour of the farmer during the rearing period dictated the catchers' success, with direct quotes including: 'if the farmer walks every day in the shed, the broilers are calmer and easier to catch' or 'if the shed is dirty, it becomes more difficult to manage'. Catchers also believe that male broilers are calmer than females, and that the catching process during rainy days is easier than during sunny days, as they perceived birds to be less agitated. In summary, catchers' seemed to have good knowledge about animal welfare and behaviour as well as about good broiler handling procedures, which were expressed by them as an important consideration to promote broilers welfare. However, a perceived lack of behavioural control regarding the conditions in the sheds and farmer management are likely barriers to optimise manual catching of broilers.

## Transporting turkey males to the slaughterhouse – does the bird size matter?

*Eija Kaukonen and Anna Valros*
*University of Helsinki, Faculty of Veterinary Medicine, Research Centre for Animal Welfare, Department of Production Animal Medicine, PL 57, 00014 Helsingin Yliopisto, Finland; eija.kaukonen@helsinki.fi*

Restricted movement may compromise turkey welfare during transportation to the slaughterhouse. In Finland, the crate height should allow birds to change their posture, lie down, stretch their legs and correct uncomfortable posture of legs and wings. This study assessed how bird dimensions and space within the crate affected turkey male behaviour during transport to the slaughterhouse. We compared males at two ages: younger males (on average 125 days, target live weight 17.0 kg) in groups of four, and older males (on average 137 days, target weight 19.7 kg) in groups of three, to keep the crate density within legal limits. Both age groups were studied in 4 transports, including 306 crates/transport. The height of the transportation crate was 40 cm and the volume 30.6 l. The volume of three randomly selected birds per slaughter batch, and the girth from 25 randomly selected birds per batch were measured at slaughter. The diameter of the birds was estimated, assuming the bird is round, using the girth measurement plus 7 cm, to account for legs, wings and feathers increasing the diameter. The body posture (laying down, sitting or low standing), unusual postures (leg pointing sideways, spread wing) and panting were recorded in the transport crates on-farm after loading, and twice in the slaughterhouse lairage after transport. Also, transport temperature was measured with data loggers at six crates per transport. Statistical analysis was conducted with Mann-Whitney U and Kruskal-Wallis tests. Younger males (n=24) had a smaller volume than older males (n=24) (18.3 l±0.52 and 21.0 l±0.43, respectively; $P=0.001$). However, the average total bird volume per crate was greater for younger males compared with older ones ($P=0.001$). Younger males (n=100) had smaller diameter than older males (n=100) (31.7±0.08 and 33.3±0.09 cm, respectively; $P=0.001$). The estimated free space between bird back and crate ceiling was thus smaller for older males than for younger males (6.7±0.09 and 8.3±0.08 cm, respectively; $P=0.001$). Older males were more often observed in an unusual posture than younger (0.6±0.28 and 2.4±0.68%; $P=0.039$). Transport temperature was higher in transports of younger males (17.3±0.07 vs 16.5±0.08 °C; $P=0.001$), but there was no difference in panting between the two age groups. Instead, panting increased after transport (on-farm 6.6±2.77% of birds were panting, at the slaughterhouse 33.3±6.44 and 34.9±9.10% were panting; $P=0.047$ and $P=0.039$, for the first and second observation respectively), indicating panting might be a sign of stress. The increased number of birds in an unusual posture at an older age, despite the lower total bird volume in the crate, could be caused by the diminished space between the crate ceiling and bird back. This suggests reduced welfare of heavier turkey males due to restricted space in vertical dimension, despite increased horizontal space.

**Relationships between type of hoof lesion and behavioural signs of lameness in Holstein cows housed in tie-stall facilities**

*Megan Jewell[1], Javier Sanchez[1], Greg Keefe[1], Michael Cockram[1], Marguerite Cameron[1], Shawn McKenna[1] and Jonathan Spears[2]*
*[1]University of Prince Edward Island, Atlantic Veterinary College, Health Management, 550 University Avenue, Charlottetown, Prince Edward Island, C1A 4P3, Canada, [2]University of Prince Edward Island, Atlantic Veterinary College, Biomedical Science, 550 University Avenue, Charlottetown, Prince Edward Island, C1A 4P3, Canada; mtrobertson@upei.ca*

Hoof lesion type and severity may or may not influence a cow to change their gait. Previous studies found that the presence of a sole ulcer (SU) was associated with changes in gait scores, whereas digital dermatitis (DD) and sole hemorrhage (SH) were not. In tie-stall facilities locomotion scoring can be difficult to perform. An alternative method, known as stall lameness scoring (SLS), allows assessments to be completed in a tie-stall to identify lameness based on behavioural changes in weight bearing and foot positioning. The aim of this study was to examine relationships between hoof lesions and SLS. Lameness was assessed by one trained observer looking for the following four behaviours: shifting weight, resting one foot, standing on the edge of the stall and uneven weight bearing when stepping side to side. Hoof trimming records were collected from two hoof trimmers trained in lesion identification. Lesion identification quizzes were used to measure the level of agreement between trimmers, both at the beginning and mid-way through the study. A high level of agreement was achieved and maintained ($\kappa > 0.80$). To ensure there was no effect of trimming on the lameness evaluation, SLS was always performed prior to trimming. Two thirds of the animals were trimmed within a week of SLS, with the most time between trimming and SLS being seven weeks. There was no significant effect of this time difference on the associations between hoof lesions and behavioural changes. The SLS behavioural changes occurred primarily in the hind limbs, therefore, the analysis was confined to hind limb lesions only. Logistic regression was used to examine associations between the presence of a hoof lesion and the SLS behavioural changes. A total of 557 cows, from seven tie-stall herds, were evaluated. At least one behavioural indicator was noted in 25% of these cows and at least one hind limb lesion was present in 19%. Of those with a lesion present, 37% were identified as DD, 34% were SU and 21% were SH. A cow with at least one lesion had a higher odds of resting one foot [odds ratio (OR)=2.96; $P<0.001$] and bearing weight unevenly when moving side to side (OR=3.75; $P<0.001$) than a cow without a lesion. Cows with a SU had a higher odds of resting one limb (OR=5.59; $P<0.001$) and bearing weight unevenly (OR=3.50; $P<0.01$) than a cow without a SU. A cow with SH had a higher odds of shifting their weight from one foot to another (OR=4.89; $P<0.01$) than those without SH and cows with DD had a higher odds of bearing weight unevenly (OR=3.63; $P<0.001$) than those without DD. Using SLS can help identify cows with hind limb hoof lesions and help producers/trimmers detect cows which may require treatment. Early intervention can help to reduce the duration of clinical lameness.

## Dairy cows alter their lying behavior when lame during the dry period

*Ruan R. Daros, Hanna Eriksson, Daniel M. Weary and Marina A.G. Von Keyserlingk*
*Animal Welfare Program, Faculty of Land and Food systems, University of British Columbia,*
*2357 Main Mall 248, V6T 1Z4, Vancouver, Canada; rrdaros@alumni.ubc.ca*

Understanding changes in behaviour may allow for early detection of lameness in dairy cows. Research to date has focused primarily on lameness and associated behavioral changes in lactating dairy cows with very little focus on the dry cow. To assess changes in lying behaviour of dry cows we conducted a longitudinal study using 306 dry cows from 6 commercial freestall dairy farms. Cows were gait scored weekly to assess lameness status (score 1 and 2 = sound, 3, 4 and 5 = lame). Lying behaviour was recorded wk -8 to -2, relative to calving, using data loggers set to record the position of the leg at 1-min intervals; loggers were attached to the hind leg of each cow and replaced bi-weekly. Lying behaviours were summarized by day as total lying time (min/d), number of bouts (no/d) and mean bout duration (min/bout). To assess changes in lying behaviours linear multilevel models were built using farm and cow within farm as random effects. The explanatory variable of interest was lameness category (sound vs lame) and wk relative to calving was used as the covariate. To specifically assess the effect of lameness on behaviour only data from cows that changed their lameness status within the study period were used. The prevalence of cows that stayed sound, lame, or changed lameness status throughout the course of the study period was 26, 24 and 50%, respectively. Regardless of lameness status, week relative to calving was associated with lying behaviour; lying time decreased by 4.9 min/wk (SEM=0.73; $P<0.01$); increased number of lying bouts by 0.35 bouts/ wk (0.01; $P<0.01$); and decreased mean duration of lying bouts by 3.6 min/wk (0.16; $P<0.01$). When lame, cows decreased their daily lying time by 12.1 min (3.03; $P<0.01$), but increased their mean bout duration by 2.4 min (0.69; $P<0.01$) compared to when sound. There was no change in the number of daily bouts when cows changed lameness status. In conclusion, lying behavior of dry cows is affected by lameness and week relative to calving. Despite the little magnitude of the behavioural changes, these results can be useful for implementation of automatic lameness detection methods.

## Effects of divergent lines on feed efficiency and physical activity on lameness and osteochondrosis in growing pigs

*Marie-Christine Meunier-Salaün[1], Maija Karhapää[2], Hilkka Siljander-Rasi[2], Emma Cantaloube[1], Lucile Archimbaud[1], Ludovic Brossard[1], Nathalie Le Floc'h[1] and Anne Boudon[1]*
*[1]PEGASE, INRA, AgroCampus Ouest, 35042 Rennes, France, [2]Natural Resources Institute Finland (Luke), Latokartanonkaari 9, 00790 Helsinki, Finland; marie-christine.salaun@inra.fr*

Locomotion disorders have been identified as a significant production disease for growing and finishing pigs, and a sign of reduced animal welfare (EU PROHEALTH project). Amongst them, lameness is a complex problem including animal-based and resources-based causes. Osteochondrosis (OC) is a very frequent disease in pigs, affecting cartilage. The study assessed the occurrence of lameness and OC in two divergent lines of Large-White selected according to a feed efficiency trait (HRFI, LRFI; high or low residual feed intake), submitted to constraints on the physical activity. Two experimental groups of 80 pigs (40 LRFI and 40 HRFI, ratio 1:1 of females and castrated males) were housed on partly slatted flooring in a room equipped with an electronic weighing device allowing pig access to electronic self-feeders, crossing a sorter. Pigs were equipped with ear tags for their individual electronic identification at the sorter and automatic feeders. The identification determined the side of the room to which the animal was directed after the sorter exit, and thus defined the distance to come back to the sorter (short: spontaneous activity, long: increased activity). Three times, at the beginning (wk3), middle (wk7) and end (wk13) of the growing-fattening period, video recordings were made during the 12 h daylight for screening the posture (standing, sitting, lying). Lameness was assessed weekly using a visual gait scoring (lameness score, 0: no, 1: lame). At slaughter, (weight of 100 kg, between 21 and 22 wks of age), post-mortem quantification of OC lesions was performed on surfaces of both proximal and distal condyles of the humerus and femur (score 0 no lesion; score 5 severe lesion with loose fragment of cartilage). According to the variables, data were analyzed using a $\chi^2$ test or variance analysis with the main factors line, physical activity, sex, replicate and their interactions. The LRFI line showed fewer pigs standing during the daylight period at the three recording times (22.5±0.6 vs 28.3±0.6% of standing pigs during a 30 min interval, $P<0.001$). The LRFI pigs also showed a higher occurrence of lameness throughout the experimental period only in the second experimental group (34/313 vs 12/299 occurrences, $\chi^2=9.16$ df=1 $P<0.01$). The scores of OC of the LRFI pigs were also higher on the proximal condyle of the humerus (2.2 vs 1.5 $P<0.001$) and femur (1.7 vs 1.2 $P<0.01$). The pigs subjected to the increased activity treatment, crossed the sorter more frequently, whatever the genetic line (5.3±0.6 vs 3.7±1.9 crossings/day for increased vs spontaneous activity, $P<0.01$), but this treatment did not affect either the number of pigs standing during the daylight period or the OC scores. For these lines, selection on residual feed intake seemed to increase the prevalence of locomotor disorders. This effect of the line has to be investigated through the analysis of behavioural strategies within and between lines.

## Osteoarthritis impairs performance in a spatial working memory task in dogs

*Melissa Smith, Jo Murrell and Mike Mendl*
*University of Bristol, School of Veterinary Science, Langford House, Langford, Bristol, BS40 5DU, United Kingdom; ms8666@bristol.ac.uk*

Osteoarthritis (OA) is a common, chronically painful condition in pet dogs. Chronic pain can impair working memory in humans, however it is not known if this is also the case in dogs. This study aimed to assess whether pet dogs with chronic pain from OA would show impaired performance in a spatial working memory task compared to healthy control dogs. 20 pet dogs with signs of OA and 21 healthy control dogs (assigned to these groups following veterinary orthopaedic examination) were recruited via posters placed in 49 veterinary practices in South West England. All dogs were neutered and between 5-10 years old. Dogs participated in a disappearing object task to assess spatial working memory, in which they were trained to touch an object in exchange for a food reward. Each dog then observed a researcher hide the object behind one of four identical boxes. An opaque barrier was then erected for an interval of 0, 60 or 120 seconds before the dog was allowed to retrieve the object. Upon visiting the correct box the dog received a food reward, but visiting an incorrect box or making no search attempt caused the dog to be led back to the starting position with no reward. This was repeated for four trials (one per box) per interval for each dog. Data were analysed using mixed-effects logistic regression in R. There was a significant effect of group on probability of successfully visiting the correct box (P(Success)) (z=-2.02, P=0.04), with osteoarthritic dogs showing a lower P(Success) (mean=0.67, 95% CI=0.61-0.73) than control dogs (mean=0.73, 95% CI=0.67-0.78). For all dogs grouped together, there was also a significant effect of the interval between observing the object being hidden and being allowed to retrieve it (z=-2.28, P=0.02), with P(Success) decreasing as the interval increased (mean P(Success) at 0 s=0.86, 95% CI=0.80-0.90; mean at 60 s=0.66, 95% CI=0.59-0.73; mean at 120 s=0.58, 95% CI=0.50-0.65). Whilst the effect of the interaction between group and interval on P(Success) was not significant (z=-1.53, P=0.13), there was a slight trend for osteoarthritic dogs' P(Success) to decrease at a greater rate than that of control dogs as the interval duration increased. This is the first study to find evidence of impaired spatial working memory associated with a chronically painful condition in dogs. Though the working memory deficits observed in this study were relatively small, it is possible that working memory deficits could potentially affect dogs' interaction with their environment and owners, which may have welfare implications for dogs with OA. Furthermore, these findings could improve the understanding of how OA and chronic pain affect cognition in dogs as well as potentially in humans and other species. This study also provides evidence supporting the use of dogs with spontaneously-developing OA as a model of human chronic pain, offering an alternative to the use of laboratory animals and experimentally-induced models of chronic pain.

## Circadian rhythms in activity in healthy and arthritic dogs

*Jack S. O'Sullivan[1], M. Cassim Ladha[1,2], Zoe Belshaw[3] and Lucy Asher[1]*
*[1]Newcastle University, Centre for Behaviour & Evolution, Institute of Neuroscience, Henry Wellcome Building, Framlington Place, Newcastle, NE4 2HH, United Kingdom, [2]VetSens, Jesmond, Newcastle, United Kingdom, [3]University of Nottingham, Centre for Evidence-based Veterinary Medicine, School of Veterinary Medicine and Science, Sutton Bonington Campus, Leicestershire, LE12 5RD, United Kingdom; j.o'sullivan4@ncl.ac.uk*

Circadian rhythms, endogenous 24-hour cyclical processes, are found in all vertebrates. Behaviour also follows a circadian rhythm which can be disrupted by poor states of welfare such as stress, anxiety, pain or disease. It is becoming easier to monitor 24-hour cycles in behaviour because modern sensor technology allows behaviour to be measured continuously over days, weeks, or even months. This means that daily rhythms in behaviour can be examined in free-living animals at resolutions and durations not previously possible. The aim of this study was to examine differences between daily rhythms of healthy dogs and those with a chronic disease. We measured circadian rhythms in the activity behaviour of pet dogs using triaxial accelerometers mounted to their collars (AX3 supplied by Vetsens). Activity data (measured using $vm^3$ from the accelerometer) were collected over 7 days in the normal home environment for 40 pet domestic dogs of various breeds: 20 dogs had been diagnosed with osteoarthritis and 20 were healthy controls. We used a cosinor mixed model to explore fit to a 24-hour circadian cycle in overall activity data. Individual dog was a random effect term in the model. Presence of circadian rhythms was explored using model fit to the cosinor model. The presence of shorter rhythms in activity was explored by sequentially adding additional harmonic terms to the model for cycles of 12, 10, 8, 6, 5, 4, 3 and 2 hours and observing improvements in model fit. Healthy dogs' activity fitted much better to a model of circadian rhythms ($R^2$=0.20) than arthritic dogs ($R^2$=0.04). Healthy dogs were less active at night compared to the day, they had two main peaks of activity during the day and the best fitting model had additional cycles of 12 and 8 hours. Dogs with osteoarthritis had lower overall activity, but appeared to be no less active during the night as compared to the day. These results suggest that arthritic dogs have a disruption in the normal circadian rhythm of activity. Findings fit with previous owner reports of frequent shifting position during rest in dogs with arthritis. There could be a number of explanations for differences between arthritic and healthy dogs in their activity patterns. The welfare implications of these different explanations will be discussed and tentative conclusions presented about the potential for circadian rhythms in general activity as a welfare indicator in dogs.

## Lameness changes cows feeding behavior

*Emma Ternman[1], Per Peetz Nielsen[2] and Lene Munksgaard[1]*
*[1]Aarhus University, Animal Science, Postboks 50, 8830 Tjele, Denmark, [2]University of Copenhagen, Veterinary and Animal Sciences, Grønnegårdsvej 8, 1870 Frederiksberg C, Denmark; emma.ternman@anis.au.dk*

Lameness is a common problem in dairy herds and has a negative effect on both animal welfare and the farm economy. To estimate the effect of lameness on cows' (n=116) behavior, data during two consecutive years were recorded and examined for relations between lameness scores, feeding behavior, activity and visits to the milking robot. Cows were scored every second week with the score 1 for non-lame cows and up to the score 5 for severely lame cows. The observations were then allocated into three lameness classes (non-lame <3, n=1,702; moderately lame = 3, n=174; and severely lame >3, n=18). Older cows (parity ≥3) had the highest lameness class (1.22±0.02) compared to parity 1 and parity 2 (1.06±0.02 and 1.07±0.02 respectively; $P<0.001$). Mid-lactation cows had lower lameness class (1.08±0.02) compared to late lactation cows (1.14±0.02; $P<0.01$). Data on feeding behavior were collected by Insentec feed system, milking by DeLaval VMS and activity by the AfiAct II activity meter. Daily means were calculated from the last 5 days before the lameness scoring was performed. The PROC MIXED procedure in SAS was used to test for significant differences between lame and non-lame cows. Lameness class, parity (1, 2, ≥3), stage of lactation (early ≤55 d, mid 56-150 d and late >150 d), and two-way interactions between these were included as fixed effects, and cow within parity as the subject of the repeated measure. Lying duration ($P=0.15$), the number of lying bouts ($P=0.46$), and the number of steps ($P=0.051$) was not affected by lameness score. Regardless of lameness class, daily lying duration was longest for cows with parity ≥3 (767.3±12.2 min) compared to cows with parity 1 and 2 (712.7±11.1 and 719.2±12.5 min, respectively; $P<0.01$). Lameness class did not affect daily milk yield ($P=0.41$) or number of visits to the VMS ($P=0.15$). The daily duration of eating for cows in early lactation decreased with increasing lameness class (non-lame 174.1±2.9 min, moderately lame 165.6±4.4 min, and severely lame 136.1±8.4 min respectively; $P<0.05$). In summary, we found that lameness negatively affected eating behavior in early lactation while no effect was found in later lactation. Lying behavior and number of steps did not differ between lame and non-lame cows.

**Prevalence of lameness based on scoring system assessing weight bearing and foot positioning of tie-stall housed dairy cattle**

*Megan Jewell[1], Javier Sanchez[1], Greg Keefe[1], Michael Cockram[1], Marguerite Cameron[1], Shawn McKenna[1] and Jonathan Spears[2]*
*[1]University of Prince Edward Island, Atlantic Veterinary College, Health Management, 550 University Avenue, Charlottetown, Prince Edward Island, C1A 4P3, Canada, [2]University of Prince Edward Island, Atlantic Veterinary College, Biomedical Science, 550 University Avenue, Charlottetown, Prince Edward Island, C1A 4P3, Canada; jsanchez@upei.ca*

Gait or locomotion scoring, looking for behavioural changes indicative of limb pain such as, back arch, head nod and uneven weight distribution, is commonly used to assess lameness, however, this can be difficult to perform when cattle are housed in tie-stalls. An alternative method, known as stall lameness scoring (SLS), was used to observe cattle in their stall for behavioural changes based on weight bearing and foot positioning. The aim of the study was to use SLS to estimate the prevalence of lameness and determine associated risk factors. Animal-, environmental- and management-based risk factors were measured on 1,511 randomly selected cows, from 34 volunteer herds, throughout the three Maritime Provinces of Canada, Nova Scotia, New Brunswick and Prince Edward Island. One observer, trained for SLS, observed the cows for 30 seconds and recorded behavioural changes of repeated shifting of weight; standing on the edge of the stall; and resting a foot. Cows were then encouraged to take steps side to side to assess for uneven weight bearing. The presence of two or more of these changes was used to categorize cows as lame. During the assessment of lameness 22% of cows were bearing weight unevenly, 14% were resting a limb, 9% were shifting weight and 6% were standing on the edge of the stall. Sixty eight percent of cows had no behavioural changes, whereas, 15, 15 and <1% of cows showed 1 to 4 behavioural changes, respectively. The prevalence of lameness was 16.7% (95% CI: 14.8-18.7%) and varied at the herd-level from 0-31%. There were numerous combinations of behavioural changes noted in cows classified as lame, with the two most common being resting a limb and uneven weight bearing (53%) and uneven weight bearing and shifting of weight (17%). A logistic regression model was used to determine risk factors significantly associated with lameness. Results indicated that the odds of lameness were lower for those bedded with forages, such as straw and hay, when compared to those bedded with other materials, such as shavings or sawdust [odds ratio (OR)=0.56; (*P*=0.020)]. When the bedding material was wet there was a higher odds of lameness compared to dry bedding (OR=1.93; (*P*=0.024)). When producers reported they had difficulty getting a hoof trimmer when needed, a higher odds of lameness was observed (OR=1.15; (*P*=0.025)). Animal-based factors such as body condition, skin lesions on extremities and age were also associated with lameness. Providing an environment with soft, dry bedding could help decrease behavioural changes associated with lameness in tie-stall facilities. Having a good relationship with a hoof trimmer to ensure cows have good hoof health and are treated promptly is important to help reduce lameness. These results could help producers to make more informed decisions to reduce the prevalence of lameness and improve the welfare of their herds.

## Rotarod – a new method for assessing the gait of broilers?

*Julia Malchow, Anissa Dudde, Jutta Berk, E. Tobias Krause and Lars Schrader*
*Institute of Animal Welfare and Animal Husbandry, Friedrich-Loeffler-Institut, Dörnbergstraße 25/27, 29223 Celle, Germany; julia.malchow@fli.de*

The gait scoring system (GSS) of Kestin *et al.* is the most frequently used method for assessing walking ability of broiler chickens by six scores ranging from 0 (normal gait, furled foot in the air) to 5 (not able to walk). Despite training of observers this method, to a certain extent, is vulnerable to subjective bias and high inter-observer variation. To quickly assess walking ability of broilers in commercial flocks, a simple, objective and reliable method is needed. Walking ability correlates to the motor coordination which can be assessed by the rotarod test, often used in pharmacological studies. We tested with broiler chickens the relationship between their gait score and the latency to leave (LTL) the rotating bar in a chickens' rotarod test. We tested 138 chickens of three strains: 46 Ross 308 (Ross, fast growing), 46 Lohmann Dual (Dual, medium growing) and 46 Lohmann Brown Plus (LB, slow growing) kept in separate 12 pens. Tests were done at a slaughter date after 5 weeks fattening period for Ross and 10 weeks for Dual and LB chickens. First, the gait score was assessed by one observer by the GSS in a test arena (0.5×2.5 m). Two conspecifics were placed at the one end of the arena to motivate the focal chicken to walk in direction of conspecifics. Directly after the GSS, chickens were gently taken to the rotarod test and sat up on the rod with both feet. After five seconds the rod started to rotate slowly by a motor and the velocity of the rotated rod increased automatically in a logarithmically course to a maximum of 2.1 sec per rotation. Latency to leave [s] the rod was measured. The rod was set at a height of 80 cm under which a box with mat at a height of 65 cm was placed to let the chickens land softly. Data were analyzed using a linear mixed model and pairwise post hoc t-tests. Gait scores, strains and weight gains were used as fixed factors and pen as random factor. Gait scores ranged from 1 to 4 in Ross, 0 to 2 in Dual, and 0 to 1 in LB chickens. LTL (n=138) were affected by gait score ($P<0.0001$), strain ($P=0.028$) and weight gain ($P=0.002$). Chickens with gait score 0 (LTL=45±20.9 s) and 1 (LTL=38.7±21.2 s) showed a significantly longer LTL compared to chickens with gait score 2 (LTL=19.1±17.7 s), 3 (LTL=12.3±10.9 s), and 4 (LTL=2.3±1.7 s) (all $P<0.01$). Ross (LTL=13.8±13.0 s) left the rotarod earlier compared with Dual (LTL=45.6±21.4 s, $P<0.0001$) and LB (LTL=44.3±20.7 s, $P<0.0001$). Furthermore, chickens with high weight gain showed a shorter LTL in contrast to chickens with lower weight gain. Especially Ross indicated difficulties to keep on the immobile rod. Results showed a significant relationship between the gait score and the LTL measured by the rotarod test. In addition, LTL was related to the weight gain. Thus, the rotarod test seems to be a valid method to more objectively measure the walking ability of chickens in a time efficient manner. Ethical and scientific animal experiment approval from LAVES.

### Epidemiological study of musculoskeletal injuries in Standardbred Racehorses on PEI

*Jamye Rouette, Michael Cockram, Javier Sanchez and Kathleen Macmillan*
*Atlantic Veterinary College, Health Management, 550 University Ave Charlottetown, C1A 4P3 PE, Canada; jrouette@upei.ca*

Musculoskeletal injuries in Standardbred horses used in harness racing are a source of pain and discomfort. They are the most important contributor to poor performance and wastage. Identification of the risk factors for injury and subsequent changes to management practices could provide benefits to the welfare of the horses and their performance. The purpose of the study was to determine the possible associations between injury and identified risk factors. A longitudinal study was designed to follow a population of Standardbred horses during their training and racing for a one-year period. A total of 136 Standardbred racehorses were enrolled and followed monthly. Information was collected from eight trainers using a questionnaire at racetracks every month. On average, each horse was studied for 4.4 months. During the study period, we observed 11 injuries over a total of 584 visits (prevalence=4%). These injuries were observed in 136 horses (prevalence=8%). Out of these 11 injuries, nine were new injuries, representing a horse level incidence of 6% for the 136 horses. The population had a total time at risk of developing new injuries of 553 months; therefore the incidence rate of injury was 0.0162 per horse-month at risk (or 1.62 cases per 100 horses in a month). New injuries were only present in pacers and no trotters experienced a new injury. The nine new injuries consisted of five bone and four soft tissue injuries that occurred while the horses were in training or racing. There were mild to severe injuries to the sesamoid and long pastern bones and damage to the suspensory ligament, superficial digital flexor tendon and deep digital flexor tendon. Depending on the type and severity of the injury determined the necessary time off, mild to moderate injuries were seen to have between three and 90 days off from training. More severe injuries required significantly more time to recover of eight months (long pastern fracture) or was career ending (sesamoid fracture). These preliminary results indicate a prevalence and incidence of injuries lower than we have anticipated. Most of the horses (87%) included in the study were between two to four years of age; the remaining were five years or older. Horses were in good condition, 97% of horses scored between 4-6 for body condition score out of a 1-9 scale. Conformation was evaluated on a score of 0-3, with zero being baseline normal and 1-3 indicating increasing severity of a conformational fault. Common conformation faults included offset knees (62% of horses) when the cannon bone is not centered directly under the carpus, camped under up front (55%) when the horse stands behind the vertical, toed out (68%) metacarpophalangeal valgus, and pastern hoof axis (75%) where pastern has an upright appearance. However, most horses with these conformation faults only had a severity score of 1 for a conformational fault. Further work is required to study horse characteristics, management and training factors that affect the risk of injuries in Standardbred Racehorses.

## Feather pecking phenotype affects behavioural responses of laying hens

*Jerine A.J. Van Der Eijk[1,2], Aart Lammers[1] and T. Bas Rodenburg[1,3]*
[1]*Wageningen University and Research, Adaptation Physiology Group, De Elst 1, 6708 WD Wageningen, the Netherlands,* [2]*Wageningen University and Research, Behavioural Ecology Group, De Elst 1, 6708 WD Wageningen, the Netherlands,* [3]*Utrecht University, Animals in Science and Society, Yalelaan 2, 3584 CM Utrecht, the Netherlands; jerine.vandereijk@wur.nl*

Feather pecking (FP) is a major welfare and economic issue in the egg production industry. It involves hens pecking and pulling at feathers of conspecifics, thereby negatively affecting welfare. Behavioural characteristics, such as fearfulness, have been related to FP. Although many studies have identified differences in fearfulness between lines that differ in FP, the relationship between actual FP behaviour (i.e. FP phenotypes) and fearfulness is not well understood. Therefore, we compared responses of birds with differing FP phenotypes to several behavioural tests at young and adult ages. We used birds from a genetic line selected for high feather pecking. FP phenotypes of individual birds were identified via FP observations at 3-4, 12-13, 15-16 and 28-29 weeks of age. The total number of severe feather pecks (SFP) given and received over two subsequent weeks was used to categorize birds as feather peckers (P, SFP given >1), feather pecker-victims (P-V, SFP given and received >1), victims (V, SFP received >1) or neutrals (N, SFP given and received 0 or 1) at each age point. Birds were tested individually in a novel environment (NE) test at 4 weeks of age, an open field (OF) test at 15 weeks of age and a tonic immobility (TI) test at 13 and 28 weeks of age. Experimenters were blinded to the phenotypes. Data were analysed using linear mixed models, with phenotype and batch as fixed factors and pen as a random factor. Test time was added as a fixed effect for the NE and OF test. Experimenter was added as a fixed effect for the NE and TI test. Testing order was included as a fixed effect for the TI test. Phenotype effects were tested for each behavioural test and age separately using the most recent FP phenotype categorization. FP phenotype affected the number of flight attempts ($F_{3,\ 119}$=3.18, $P<0.05$) during the NE test, where victims showed more flight attempts compared to neutrals (V=2.3 vs n=1.6; $P<0.05$) and tended to show fewer flight attempts compared to feather peckers (V=2.3 vs P=2.7; $P<0.1$). FP phenotype further tended to affect step frequency ($F_{3,\ 75}$=2.64, $P<0.1$) during the OF test, where feather peckers tended to walk more compared to neutrals (P=24.6 vs n=15.7; $P<0.1$). No FP phenotype effects were found for the TI test. Feather peckers tended to show more active responses (i.e. tended to show more flight attempts compared to victims and tended to walk more compared to neutrals), which could suggest lower fearfulness, compared to victims at 4 weeks of age and compared to neutrals at 15 weeks of age. These findings give first indications that FP phenotypes seem to differ in fearfulness. It should be noted that we only found differences in the NE and OF test, where behavioural responses could also be related to activity or coping style. Further research is needed to identify whether FP phenotypes differ in activity and whether they can be classified into different coping styles.

## An investigation of associations between management and feather damage in Canadian laying hens housed in furnished cages

*Caitlin Decina[1], Olaf Berke[1], Christine F. Baes[2], Nienke Van Staaveren[2], Tina Widowski[2] and Alexandra Harlander-Matauschek[2]*
[1]*University of Guelph, Department of Population Medicine, Ontario Veterinary College, 50 Stone Road E, Guelph, Ontario N1G 2W1, Canada,* [2]*University of Guelph, Department of Animal Biosciences, Ontario Agricultural College, 50 Stone Road E, Guelph, Ontario N1G 2W1, Canada; cdecina@uoguelph.ca*

Canada is in the process of transitioning from conventional cages to alternative housing systems for laying hens, such as furnished cages, equipped/furnished with perches, nest areas and scratch pads, to improve hen welfare. However, feather damage (FD) due to feather pecking remains a welfare concern in furnished cages. With the goal of achieving the desired improvement of hen welfare in practice, we took an explorative approach to assessing bird, housing and management associations with FD in laying hens in furnished cage and non-cage systems in Canada. A questionnaire, focused on housing and management practices, was designed and administered to 122 laying farms nation-wide in autumn of 2017 (response rate of 52.5%), providing information on a total of 65 flocks housed in alternative systems (26 in furnished cages). In conjunction, a simple, three-point feather cover scoring system (0: fully feathered; 2: poorly feathered with a naked area larger than a 2-dollar coin) was developed for farmers to estimate the prevalence of FD present on-farm. Farmers visually assessed a sample of 50 birds obtained evenly throughout the barn, providing a FD score for the back/rump area. Flock prevalence of FD was determined by the percentage of birds with a score greater than zero. Linear regression modelling in R was used in a preliminary analysis of the effect of 30 potential variables related to bird characteristics, nutritional aspects and general management on the prevalence of FD. Variables were screened in a univariable analysis ($P<0.25$) where 14 remained for multivariable modeling. Six predictors were included in the final model: age (centered at 40 weeks), feather colour, frequency of feeder running, frequency of manure belt running, number of workers performing daily inspections, and provision of a scratch area. FD prevalence was estimated at ~60% at 40 weeks of age within the model, and had a positive association with age, suggesting FD increases over time. White feathered birds were associated with lower prevalence of FD compared to brown-feathered birds. More frequent feeder runs were positively associated with slightly higher FD. In contrast, more frequent use of the manure belt was associated with substantially lower FD. Finally, lower FD was also observed when a scratch area was provided or when 3 or more workers performed daily inspections. The predictors and their subject areas investigated have previously been associated with FD, as they are suggested to influence stress/fear in laying hens (e.g. genetic factors, behavioural restrictions, environmental disturbances/unfamiliarity). Further longitudinal investigations into management practices in furnished cages could better elucidate the associations found in this study and help farmers improve laying hen welfare during the transition to alternative systems.

## Enriching floor pens mitigates effects of extended pullet housing on subsequent distribution of laying hens in an aviary

*Sarah S. Maclachlan[1], Ahmed B. Ali[1], Michael Toscano[2] and Janice M. Siegford[1]*
[1]*Michigan State University, Animal Science, 474 S Shaw Ln, East Lansing, MI 48824, USA,*
[2]*VPH-Institute, Universiry of Bern, Division of Animal Welfare, Hochschulstrasse 6, 3012 Bern, Switzerland; sarahsmaclachlan@gmail.com*

Aviaries provide hens with many resources, but the birds must develop motor and cognitive skills to be able to use them properly. In production, pullets are typically moved into layer housing at ~17 weeks of age as they start to lay; however delayed movement of pullets to layer facilities can occur for a variety of reasons. Introducing birds to the aviary at a later age could result in less use of perches, nests, and vertical space, which could impact both productivity and welfare of the birds. Of particular concern, hens may lay more eggs on the floor if not introduced to aviary nests as they start to lay, and they may be more likely to injure themselves when moving through the complex space. The objectives of this study were: (1) to examine flock level distribution of hens experiencing extended pullet housing; and (2) to examine how enrichment influences the distribution of birds experiencing extended pullet housing. In this study, 3 treatments were applied with 4 aviary units/treatment and 100 Hy-Line W36 laying hens/unit. Pullets were raised in floor pens and placed into aviaries at 17 weeks (Control, CON) and 25 weeks (Floor, FLR and Enriched, ENR). Additionally, ENR hens were provided with perches and nest boxes at 17 weeks of age. The number of birds in each level of the aviary (litter, lowest tier, middle tier, and highest tier – including the nest) was counted twice at each of three times of day (morning, midday, and evening), and over two periods of lay (peak and mid). The data were analyzed using generalized linear mixed models in R statistical software. In the morning period at peak lay, ENR and CON birds occupied areas in the aviary similarly and differently from FLR birds. For example, on average, 34.25% of CON hens and 30.25% of ENR hens occupied the highest tier, while only 14.78% of FLR hens were observed in the highest tier ($P<0.01$). In the midday period during peak lay, similar numbers of CON (57.25%) and ENR birds (56.5%) were seen in the litter compared with 76.63% of FLR birds ($P<0.01$). During peak lay, CON and ENR hens laid fewer eggs on litter than FLR hens (19.35 and 26.43, respectively vs 51.06%, $P<0.01$). While during mid lay the difference in floor laying between FLR and CON/ENR hens was less pronounced (CON: 6.91%, ENR: 8.75%: FLR: 23.39, $P<0.01$). Finally, in the evening during peak lay, CON and ENR hens moved to the highest tier of the aviary in greater numbers than FLR hens (21.6 and 17.2%, respectively vs 7.2%, $P<0.01$). In the mid lay period, hens followed similar patterns of distribution, but differences between FLR hens and CON/ENR hens were less dramatic, suggesting FLR hens were adapting to the aviary. Overall, we conclude that supplementing pullet floor housing with perches and nest boxes appears to help birds adjust to an aviary if they are moved in at older ages.

## Enriching floor pens mitigates effects of extended pullet housing on resource use and activity of individual hens in aviaries

*Ahmed B. Ali[1], Michael Toscano[2] and Janice M. Siegford[1]*
[1] *Michigan State University, Department of Animal Science, 474 S Shaw Ln Room 1290, East Lansing, MI 48824, USA,* [2]*VPH-Institute, University of Bern, Division of Animal Welfare, Hochschulstrasse, 3012 Bern, Switzerland; siegford@msu.edu*

Pullets are typically moved into aviaries before they have begun to lay (~17 wk of age). However, extended pullet housing may be necessary in certain situations, causing birds to first interact with complex aviary configurations at older ages. This delayed move might impact hens' ability to use and navigate through aviary systems. The objectives of this study were to investigate: (1) possible impacts of extended pullet housing on hens' ability to navigate through and use various resources in an aviary system; and (2) whether enriching pullet housing with perches and nests would mitigate such impacts. Three treatments were applied to 4 aviary units/ treatment (100 Hy-Line W36 laying hens/unit). Control (CON) pullets were raised in floor pens and placed into aviaries at 17 wk old. Floor (FLR) pullets were raised in floor pens and placed into aviaries at 25 wk old. Enriched (ENR) pullets were raised in floor pens and given perches and nests at 17 wk, then placed into aviaries at 25 wk old. Individual hen tracking was conducted on 5 randomly selected hens/unit (n=20 hens/treatment) using a different colored vest containing a triaxial accelerometer (Onset HOBO Pendant®G, scanning frequency of 20 Hz ± 3 g). Direct observations were performed for 3 min/hen in Morning, Midday and Evening for 3 consecutive days at Peaklay (36 wk of age) and Midlay (54 wk of age), while loggers recorded hens' activity across the 3 days of each period. Duration of time spent in each resource (litter area and the 3 vertical tiers) was calculated as % of total time observed for each hen. Hen's total triaxial acceleration (g) and acceleration on dorsoventral (vertical activity among tiers) and craniocaudal axes (horizontal activity within tiers) during light hours and frequency (n), distance (cm), and force of collision (N) of falls during dark hours were measured by data loggers. Data were analyzed using GLMM with treatment, time of day and stage of production as fixed effects and unit and day of observation as random effects (α set at 0.05) with R software. Acceleration data were handled using MATLAB. During Peaklay, FLR hens spent more time on litter (74 vs 58% CON or 44% ENR, $P$=0.01), while ENR hen spent more time in the highest tier (21 vs 11% CON or 5% FLR, $P$=0.02). Triaxial activity was not different among treatments; however, activity among tiers was higher in ENR (0.8 g) than CON (0.6 g) and FLR hens (0.3 g; $P$=0.003). CON hens showed the highest fall incidence (16 vs 9 FLR and 5 ENR), while FLR hens showed the highest collision force and acceleration (15 N and 0.8 g vs 10 N and 0.5 g in CON and 5 N and 0.3 g in ENR hens; all $P$≤0.05). CON and ENR hens fell from higher locations than FLR hens, with average distances of 151, 148 vs 102 cm, respectively (all $P$≤0.05). During Midlay, findings from ENR and CON hens were very similar, while results of FLR hens were similar to those of CON hens. Extended pullet housing in floor pens impacts hens' subsequent resource use, activity and falls during nighttime. However, enriching pullets' housing considerably mitigated such impacts.

## Too close for comfort – effects of stocking density on the comfort behaviours of three strains of pullets in enriched cages

*Lorri Jensen and Tina Widowski*
*University of Guelph, Animal Bioscience, 50 Stone Road East, N1G 2W1, Canada;*
*ljensen@uoguelph.ca*

In 2017, Canada was among the first countries to specify space requirements for pullets (271 $cm^2$/bird). However, little is known about the effects of different stocking densities (SD) on pullet behaviour, particularly when perches are provided. This study focused on the effects of SD on the frequency of comfort behaviours such as wing flaps, leg stretches, feather raises, and body shakes performed by pullets in closed rearing cages (Combi Pullet, Farmer Automatic). These cages were furnished with 3, 238 cm perches and a raised platform running the length of the cage. Floor space allowance ($cm^2$/bird, including platform) and number of birds per cage were 246.5 $cm^2$, 91 birds; 270.3 $cm^2$, 83 birds; 298.7 $cm^2$, 75 birds; and 334.8 $cm^2$, 67 birds. Two white strains (LSL-lite, Dekalb) and 1 brown strain (Lohmann Brown[LB]) were used, and there were 3-4 replicate cages per strain and SD (n=45). Cages were located in 2 rooms and balanced for strain and SD. Video recording was not possible due to obstructions from perches and difficulty seeing the back of the cage, so researchers performed live observations of the pullets in their home cages at 14 wk of age for 20 min per cage, 3 times per day (after lights on, midday, and prior to lights off). Focal animals were selected by choosing the bird closest to certain points in the cage at the start of an observation. Five start locations were on the cage floor, 2 on perches, and 1 on the raised platform. These reflected the observed levels of birds perching (19.5%±0.5) and using the platform (9.2%±0.1). During a 20 min observation, a focal bird was selected according to a pre-assigned start location and observed continuously for 10 min. This was repeated for a second focal with another starting location. Occurrences of wing flaps, leg stretches, feather raises, and body shakes were recorded. We hypothesized that the largest strain (LB) housed at the highest SD would perform the fewest comfort behaviours. A generalized linear mixed model was used to evaluate the effects of strain, SD, time of day, and start location on the sum of all comfort behaviours. Overall, the pullets performed an average of 0.94±0.09 comfort behaviours per 10 min observation. Feather raises occurred most frequently (0.39±0.05) and wing-flaps the least (0.14±0.03). SD and time of day did not affect comfort behaviours ($P=0.851$ and $P=0.235$, respectively). LBs performed the fewest comfort behaviours and LSLs performed the most (strain; $P=0.049$). There was no interaction between strain and SD ($P=0.537$). Start location was significant ($P<0.001$) with birds performing 3 times more comfort behaviours on the perches and raised platform than on the cage floor. This suggests that pullets may seek out the necessary space to perform comfort behaviours on perches or platforms, away from the commotion of the floor. Thus the provision of vertical spaces away from feeders or other busy resources may give more birds the opportunity to perform comfort behaviours.

### The use of pecking pans to reduce injurious pecking and enhance welfare in young pullets

*Paula Baker[1], Christine Nicol[2] and Claire Weeks[1]*
[1]*University of Bristol, Bristol Veterinary School, Langford House, Langford, Bristol, BS40 5DU, United Kingdom,* [2]*Royal Veterinary College, Royal College Street, London, NW1 0TU, United Kingdom; p.e.baker@bristol.ac.uk*

All systems of egg production generate welfare challenges. One of the most problematic issues in commercial egg production is injurious pecking (IP). IP is identified by the pecking or pulling of the feathers or flesh of another flock member, which can cause distress, feather loss or even death. As IP is thought to develop from thwarted foraging, one approach to managing it is to improve or 'enrich' the birds' environment. This study investigated the effect on feather cover and injurious pecking behaviour of provision of pecking pans during rear. Sixteen commercial flocks were used: British Blacktail (n=8), Lohmann Brown (n=6) and Bovan Browns (n=2). Flock size ranged from 3,300-11,000 (mean 6,843). Half of the 16 flocks were beak trimmed (BT) at a day old at the hatchery and the other half were intact beak flocks. Pecking pans were supplied by 6 weeks of age to 4 BT and 4 intact beak flocks to give a 4×4 experimental design. Each flock was visited three times throughout the rearing period for observations of IP behaviour and use of the pans at 6-7 weeks of age 10-11 weeks and a final visit at 14-15 weeks of age when plumage was also scored on 153 birds per flock. 5 minutes scan observations of IP and aggressive behaviour during the same observation were recorded in 9 areas of the house, (total of 45 minutes each visit) to give a representative sample. Focal sampling techniques were used to record usage of the pecking pan in treatment flocks. 2 individual birds were then counted pecking (per peck) at the pecking pan for up to 2 minutes (4 minutes in total). There was a reduction with age in pecking pan activity for the 8 flocks provided with pecking pans with pecking activity seen more frequently in birds at the 1st visit compared to visits 2 and 3 (Visit 1 $\chi^2$(2)=75.69, df 2, *P*<0.001). Intact beak birds used the pecking pan more than beak-trimmed birds over all 3 visits Mann Whitney (U=12,43, median (Md)=0.3658, n0=150; n1=192, *P*<0.31). Intact flocks pecked at the pecking pan substrate more at 14-15 weeks of age (U=1.11, Md=0.3321 n0=50, n1=64, *P*<0.006) compared to beak trimmed flocks. There was no significant difference in observations of IP behaviour in either beak trimmed or intact beak birds with or without access to pecking pans throughout the three visits and neither were any breed effects seen. A reduction of plumage damage in some body areas was seen at 14-15 weeks of age in flocks with access to pecking pan. Mann Whitney test revealed less tail plumage damage (U=53.86, Md=0.00, n0=330; n1=360, *P*<0.015) and wing plumage damage (U=57.16, Md=0.00 n0=330; n1=360, *P*<0.005). However, there was no significant difference in plumage condition for neck, back or rump regions in birds with or without a pecking pan. This study demonstrates that pecking pans for chicks/pullets can provide a long-lasting environment enrichment which may reduce plumage damage and enhance bird welfare.

## Animal care by Canadian farmers in relation to feather cover damage in laying hens

*Nienke Van Staaveren[1], Caitlin Decina[2], Christine F. Baes[1], Tina Widowski[1], Olaf Berke[2] and Alexandra Harlander-Matauschek[1]*
[1]*University of Guelph, Department of Animal Biosciences, Ontario Agricultural College, 50 Stone Road E, Guelph, Ontario N1G 2W1, Canada,* [2]*University of Guelph, Department of Population Medicine, Ontario Veterinary College, 50 Stone Road E, Guelph, Ontario N1G 2W1, Canada; nvanstaa@uoguelph.ca*

Much emphasis is placed on housing of laying hens and how this affects their welfare, leading to changes such as the phasing out of conventional cages in Canada. Feather cover damage (FD) due to feather pecking is a problem in all housing systems, but can be difficult to manage in furnished cages or non-cage systems where birds have access to a high number of potential pecking victims. However, regardless of housing, farmers can significantly affect laying hen welfare through the way they perform routine animal care. A better understanding of how Canadian farmers perform routine care and how this is associated with FD can help develop management tools to prevent FD. A questionnaire was designed and sent out to a total of 122 laying hen farmers with alternative housing systems. Farmers were asked a variety of questions regarding their flock, housing and management practices, however, here we focus on questions related to farmer characteristics and animal care, specifically. In addition, farmers received visual aids and detailed instructions to assess 50 birds for FD on the back/rump area using proportionate stratified sampling to ensure representation of the entire flock. A 3-point scale was used where 0: fully feathered and 2: naked spots larger than a two-dollar coin ($\pm2.5$ cm diameter). The percentage of birds with a score greater than zero was calculated to determine the prevalence of FD in each flock. Generalized linear mixed models were used to investigate FD prevalence and its association with variables related to animal care. Variables which lacked variation or were related to each other were excluded, leaving characteristics of the farmer such as work experience, number and duration of inspections, and whether they look for signs of FD and feather eating to be included. Sixty-four questionnaires were returned (52.5% response rate) providing useable information on 65 flocks about FD and animal care. The average prevalence of FD was 24.3% (95%CI 16.81-31.81%) with average flock age of 45 wks (95%CI 41.5-48.7 wks). Flocks were both brown (50%) and white feathered (50%) representing the majority of breeds commonly used in Canada. The following practices were associated with a higher prevalence of FD: farmers spending less than 30 min on daily bird inspection (+18.8%, 95%CI 2.35-35.33%) and if feather eating from the floor had been observed by farmers in the current flock (+27.6%, 95%CI 3.97-51.24%). Farmers not looking for signs of FD during inspections also tended to be associated with a higher FD prevalence (+14.7%, 95%CI -0.40-29.76%). Work experience of the farmer with laying hens and number of daily inspections of the birds (which was unrelated to the duration of the inspection) were not associated with the prevalence of FD. These results highlight the importance of a good and observant animal care routine and awareness of FD as a way to monitor and reduce/prevent FD.

## Development of furnished cages from conventional cages: which do hens like a nest mat or a litter mat for the nest area?

*Atsuko Kikuchi, Katsuji Uetake and Toshio Tanaka*
*Azabu University, Graduate school of veterinary science, 1-17-71 Fuchinobe, Chuo-ku, Sagamihara-shi, Kanagawa, 252-5201, Japan; da1601@azabu-u.ac.jp*

Because laying hens are mainly kept in conventional cages in Japan, the aim of the study was to develop housing systems which are easy to modify from existing systems and low in cost. Six conventional cages were combined into one cage (144 cm width, 35 cm depth, and 41 cm height at the rear). Two wooden perches (13.7 cm/hen), claw sharpeners, feed troughs, and drinkers were provided. Both ends of the cage were partitioned by curtains as nest areas (24 cm width, 35 cm depth), then there were 2 nest areas in a cage. Nest areas were combined with a dust bath. Pullets were introduced to the modified cage when they were 17 weeks. Our previous study suggested that nest areas with the artificial turf were used more than the bare cage. In this study, we evaluated the effects of meshed nest mat and plastic litter mat used in the Eurovent EU cage system on the use of the nest areas. The entire nest areas of 4 cages were lined with litter mats, and those of the other 4 cages were lined with nest mats. Fifty-six White Leghorn (Julia) hens aged 32 weeks were allocated at 7 birds per cage with a space allowance of 720 cm$^2$/hen. Direct visual scans at 10-minute intervals were conducted 4 h/d, 3 d/wk for 4 weeks (total 12 days) to count the behaviors and locations of hens. Behaviors recorded were: eating and drinking, resting, comfort behaviors (body shaking, wing flapping, tail wagging, preening, head scratching, beak wiping, and stretching), sham dust bathing, exploring, aggressive pecking, and others. The locations were recorded as nest area, perch, and floor. Where eggs were laid was also recorded. The proportions of each behavior and location were calculated for each cage for 12 days. The effects of mats on the proportions of each behavior and location were analyzed using Mann-Whitney's U-tests. The effects of the mats on the location of sham dust bathing and egg laying were analyzed using $\chi^2$ tests. Hens were observed in the nest areas more frequently in cages with nest mats compared to those in cages with litter mats (26.4 vs 20.9% of hens, respectively; $P<0.05$). Hens in cages with the nest mats were observed in the nest areas significantly more frequently over other areas for performing sham dust bathing (47.9 vs 52.1% of hens, respectively, expected value: 33.3%; $P<0.01$). On the other hand, hens in cages with the litter mats were observed in other areas significantly more often than nest areas for performing sham dust bathing (88.7% of hens vs 11.3% of hens, respectively, expected value: 33.3%; $P<0.01$). Significantly more eggs were laid in nest areas than the other areas in both cages (50.9% of eggs on the litter mats vs 53.6% of eggs on the nest mats, expected value: 33.3%; $P<0.01$). In conclusion, the nest mat was used more than the litter mat for sham dust bathing and nesting. The results suggest that the nest mats were effective for the nest area which were combined with a dust bath in the modified cage.

## Development of furnished cages from conventional cages for laying hens: square cage design and comparison of nest mats

*Toshio Tanaka, Atsuko Kikuchi and Katsuji Uetake*
*Azabu University, Graduate school of veterinary science, 1-17-71 Fuchinobe, Chuo-ku, Sagamihara, Kanagawa, 252-5201, Japan; tanakat@azabu-ac.jp*

The aim of this study was to develop a furnished cage using the conventional cages that are mainly used in Japan. In this study, a cage design and effects of nest mats were evaluated by observing behaviors of hens and their use of resources. Six conventional cages were bound together to make a square furnished cage (72 cm width, 70 cm depth, 41 cm height) equipped with 2 perches (13.7 cm/hen), claw sharpeners, and feed and water troughs divided on both side of the cage. The light schedule was 14 h light (06:00 to 20:00). Fifty-six White Leghorn (Julia) hens aged 43 weeks were divided among the 8 furnished cages. The area per hen (720 $cm^2$/hen) was less than that in EU regulations (750 $cm^2$/hen) and nest areas were combined with a dust bath due to restrictions in the design of the battery cages. Curtains partitioned 2 nest areas (24 cm width, 35 cm depth) per cage. The nest areas of 4 cages were all lined with polyethylene artificial turf, and the nest areas of the other 4 cages were all lined with commercial nest mats for the Eurovent EU cage system. Behavioral observations were conducted at 10-min intervals using a direct visual scan technique for 4 h/d (10:00 to 12:00, 13:00 to 15:00), 3 d/wk for 4 weeks (total 12 days). Behaviors observed were: eating and drinking, resting, comfort behavior (body shaking, wing flapping, tail wagging, preening, head scratching, beak wiping, and stretching), sham dust bathing, exploring, aggressive pecking, and other behaviors. The locations were recorded as nest area, perch, and floor. The places where eggs were laid were recorded. Each behavior and its location in each cage are presented as percentages of the total observations. The effects of mats on each behavior and use of resources were analyzed using Mann-Whitney's U-tests. Use of the nest areas was analyzed using $\chi^2$ tests. The expected values were calculated from the ratio of 1 to 2 for the dimensions of the nest areas to other areas. There were no significant differences on proportions of each behavior and location between types of nest mats. Hens were observed to perform sham dust bathing in nest areas with the artificial turf (50.7%, expected value: 33.3%; P<0.01). There was a tendency for hens in cages with the commercial nest mats to perform sham dust bathing in other areas (74.2%, expected value: 66.6%; P=0.05). Significantly more eggs were laid in the nest areas than in other areas in both cages (artificial turf: 59.8% vs commercial nest mat: 52.1%, expected value: 33.3%; P<0.01). In conclusion, artificial turf was used by some hens in the square furnished cage for both dust bathing and nesting. The square design cage which had nest areas with a combined dust bath with artificial turf was an effective modification of a conventional cage.

## The assessment and alleviation of pain in pigs – where are we now and what does the future hold?

*Sarah Ison[1], Eddie Clutton[2], Carol Thompson[3], Pierpaolo Di Giminiani[4] and Kenny Rutherford[3]*
*[1]World Animal Protection, 5th Floor, 222 Grays Inn Road, London WC1 X8HB, United Kingdom, [2]University of Edinburgh, Royal (Dick) School of Veterinary Studies, Easter Bush Veterinary Centre, Roslin EH25 9RG, United Kingdom, [3]Scotland's Rural College (SRUC), Animal behaviour & welfare, Roslin Institute Building, Easter Bush, Roslin EH25 9 RG, United Kingdom, [4]Newcastle University, School of Agriculture, Agriculture Building, Newcastle upon Tyne NE1 7RU, United Kingdom; sarahison@worldanimalprotection.org*

When non-human animals are used for human benefit, there is a moral obligation to prevent or minimize pain. Paradoxically, whilst animal models have benefitted the understanding of pain mechanisms in humans leading to advances in human pain medicine, the capacity of animals to experience pain is a relatively recent subject of scientific interest. One reason for this is the difficulty in assessing pain. In humans, verbal self-report is the 'gold standard' for pain assessment, but even this approach is subjective, and no 'gold standard' exists for non-human animals. The lack of simple and objective pain assessment tools makes it difficult to evaluate the welfare consequences of putatively painful states and hampers the development of pain mitigation strategies. The pig has become the subject of an increasing volume of pain research because of: (1) the prevalence of presumably painful management procedures (e.g. castration, tail-docking) conducted on-farm; (2) interest in the welfare significance of clinical conditions (e.g. lameness); and (3) the burgeoning use of laboratory pigs as human research models. This interest in the experience of pain in pigs has led to societal concern over painful interventions (e.g. husbandry procedures), has provided a platform to approve and implement pain management strategies, and provided alternatives to end the use of painful interventions. This presentation will address the challenges of studying pain in pigs, particularly the affective component, using the increasing body of scientific evidence to describe this subjective, personal experience. New insights into the pig pain experience and the implications of these for pig management and welfare will be discussed. Finally, barriers to the implementation of pain mitigation both in relation to replacing painful procedures with alternatives or providing pain relief will be discussed, using this evidence to consider how the lives of millions of pigs may be improved by wider stakeholder engagement. The presentation will incorporate scientific research and wider social science perspectives to provide novel insights into pain as a pig welfare issue.

## A multimodal approach for piglet pain management after surgical castration

*Abbie V. Viscardi and Patricia V. Turner*
*University of Guelph, Pathobiology, 50 Stone Road E, Guelph, Ontario, N1G 2W1, Canada;*
*aviscard@uoguelph.ca*

Millions of piglets in North America and the EU are surgically castrated each year, to prevent boar taint and reduce aggression. Nonsteroidal anti-inflammatory drugs (NSAIDs) are recommended for use on-farm to manage pain; however, recent research has determined that administering a single dose of meloxicam (0.4 mg/kg or 1.0 mg/kg) or ketoprofen (6.0 mg/kg) were ineffective at alleviating castration pain in piglets. Previous work in our lab found 0.04 mg/kg buprenorphine, administered alone, was highly effective at managing piglet castration pain without causing any obvious side effects. In companion animal medicine, animals are often provided with an NSAID, opioid, and anesthetic to manage peri-operative pain. The aim of this study was to assess the efficacy of 0.4 mg/kg meloxicam (MEL), 0.04 mg/kg buprenorphine (BUP), and/or Maxilene (MAX) in managing piglet castration pain, using behaviour, vocalization, and facial grimace analysis. Five-day-old male piglets from 25 different litters (n=150 total, 15 per treatment group) were randomly assigned to one of the following treatments: MEL + BUP + MAX (castrated or uncastrated), MEL + BUP (castrated or uncastrated), BUP + MAX (castrated or uncastrated), MEL + MAX (castrated or uncastrated), saline (castrated control), or sham (uncastrated control). Treatments were administered intramuscularly (MEL, BUP, saline) or topically on the scrotal surface (MAX) 20 min prior to surgical castration. Piglets were video recorded 1 h pre-procedure, immediately post-castration for 8 h and for another hour at 24 h post-procedure. Behaviour was scored continuously for the first 15 min of every hour using Observer XT software and analysed using a GLIMMIX model with post-hoc Tukey-Kramer adjustment. Still-images (n=1,118) of piglet faces were captured from the first 30 min of each hour and scored using the Piglet Grimace Scale (PGS). Vocalizations were recorded from each piglet at three points in the study: when they were handled, injected, and castrated. Both PGS and vocalization results were analysed with a mixed model procedure and post-hoc Tukey test. Castrated piglets in the MEL + BUP + MAX, MEL + BUP, and BUP + MAX treatment groups displayed significantly fewer pain behaviours (tail wagging, spasms, trembling, stiffness, rump scratching) than piglets administered saline ($P<0.0001$). Compared to saline-treated piglets, MEL + MAX was insufficient in reducing surgical castration pain behaviours ($P=0.1269$). At 24 h post-procedure, saline and MEL + MAX- castrated piglets displayed significantly more pain behaviours than all other treatment groups and time points ($P<0.01$). PGS scoring indicated that MEL + MAX-castrated piglets grimaced significantly more than MEL + BUP (castrated and uncastrated) and BUP + MAX- uncastrated ($P<0.05$). There were no significant differences in emitted vocalizations between the analgesia-treatment groups and saline ($P>0.05$). All treatment groups with BUP were successful in alleviating castration-associated pain behaviours, suggesting that opioid administration is highly effective for managing piglet castration pain.

**A systematic review of local anesthetic and/or systemic analgesia on pain associated with cautery disbudding in calves**

*Charlotte B. Winder[1], Cynthia L. Miltenburg[1], Jan M. Sargeant[1,2], Stephen J. Leblanc[1], Derek B. Haley[1], Kerry D. Lissemore[1], M. Ann Godkin[3] and Todd F. Duffield[1]*
*[1]University of Guelph, Population Medicine, 50 Stone Rd. E, Guelph, Ontario, N1G 2W1, Canada, [2]University of Guelph, Center for Public Health and Zoonoses, 50 Stone Rd. E, Guelph, Ontario, N1G 2W1, Canada, [3]Ontario Ministry of Agriculture, Food, and Rural Affairs, 6484 Wellington Rd. 7, Unit 10, Elora, Ontario, N0B 1S0, Canada; winderc@uoguelph.ca*

Disbudding is a common management procedure performed on dairy farms, and when done without pain mitigation is viewed as a key welfare issue. Use of pain control has increased in recent years, but full adoption by veterinarians or dairy producers has not been achieved. This may in part be due to the lack of consistent recommendations of treatment protocols in studies examining pain control methods for disbudding. The objective of this systematic review was to examine the effects of these pain control practices for cautery disbudding on outcomes associated with disbudding pain in calves: plasma cortisol concentrations, pressure sensitivity of the horn bud area, and pain behaviours (ear flick, head shake, head rub, foot stamp, and vocalization). Intervention studies describing cautery disbudding in calves 12 weeks of age or younger were eligible, provided they compared local anesthesia, non-steroidal anti-inflammatory drug (NSAID), or local anesthesia and NSAID, to one or more of local anesthesia, NSAID, or no pain control. The search strategy used the Agricola, Medline, and Web of Science databases, as well as the Searchable Proceedings of Animal Conferences, ProQuest Dissertations and Theses, and Open Access Theses and Dissertations. Meta-analysis was performed for all outcomes measured at similar time points with more than two studies. Local anesthetic was associated with reduced plasma cortisol until 2 h post-disbudding; however, a rise in cortisol was observed in the meta-analysis of studies reporting at 4 h post-disbudding. Heterogeneity was present in several of the analyses for this comparison. The addition of NSAID to local anesthetic showed reduction in plasma cortisol at 4 h, and a reduction in pressure sensitivity and pain behaviours in some analyses between 3 and 6 h post-disbudding. Heterogeneity was present in some meta-analyses, including several using pain behaviour outcomes. This may reflect the variation in measurement time periods for behavioural measures between studies, as well as differences among NSAID treatments. Overall, a protective effect of local anesthetic was seen for the acute pain of cautery disbudding, and the delayed rise in cortisol was mitigated by the addition of an NSAID, which also reduced other signs of pain including pressure sensitivity and pain behaviours. Based on these findings, we recommend use of local anesthetic and an NSAID as best practices for pain mitigation for cautery disbudding of calves 12 weeks of age or less. The magnitude and duration of the effect of NSAID treatment was not possible to deduce from the literature because there was much variation between studies. We recommend consideration of more standardized outcome measurements, especially for pain behaviours. Adherence to reporting guidelines by authors would help ensure more transparent and complete information is available to end users.

## Do dairy calves experience ongoing, non-evoked pain after disbudding?

*Sarah J.J. Adcock and Cassandra B. Tucker*
*Center for Animal Welfare, University of California, 1 Shields Ave, Davis, CA 95616, USA;*
*sadcock@ucdavis.edu*

Hot-iron disbudding, a husbandry procedure that prevents horn bud growth through tissue cauterization, is painful for calves. The resulting burns remain sensitive to mechanical stimulation for weeks, but it is unknown whether calves experience ongoing, non-evoked pain. We assessed whether a conditioned place preference paradigm could detect ongoing pain in the immediate hours, as well as 3 weeks, after disbudding. We evaluated conditioned place preference for analgesia in 44 calves disbudded or sham-disbudded 6 hours (Day 0) or 20 days (Day 20) before testing (n=11/treatment). Calves had 6 conditioning trials over 3 days to pair the effects of an injection of lidocaine or control (saline) with the location and pattern of a visual stimulus. For calves tested on Day 0, we disbudded the left horn bud before conditioning to make the association with pain relief salient. Thus, during each conditioning trial, Day 0 calves received an injection on the left side only, while Day 20 calves received injections on both sides. Next, on the $4^{th}$ day, we disbudded the right bud for Day 0 calves and assessed all animals' preference for the lidocaine- vs saline-paired stimulus over a 5-min period. We also measured mechanical nociceptive thresholds (MNT) around the burns before conditioning and after the preference test. On Day 0, disbudded calves tended to prefer the lidocaine-paired stimulus (Wilcox test: median proportion of time (IQR): 0.62 (0.11); $P=0.083$), while sham calves showed no preference (0.62 (0.61), $P=1$), providing evidence that disbudded calves tended to find analgesia rewarding. MNT was lower in disbudded calves (general linear model; mean ± SE: Disbudded: 1.15±0.20 N; Sham: 2.35±0.21 N; $P<0.001$), indicating evoked pain was present on Day 0. On Day 20, sham calves avoided the lidocaine-paired stimulus (0.27 (0.46), $P=0.045$), while the disbudded calves showed no preference (0.57 (0.22), $P=0.240$). MNT beforehand tended to be lower in disbudded calves (Disbudded: 1.86±0.45 N; Sham: 3.07±0.45 N; $P=0.070$), but then subsequently decreased in sham calves ($P=0.030$) such that there was no difference after testing (Disbudded: 1.70±0.27 N; Sham: 1.94±0.27 N; $P=0.524$). The sham calves' aversion to lidocaine and decrease in MNT suggests that the drug was painful. Disbudded calves did not show conditioned place aversion, possibly because they were willing to make a trade-off between the pain of the lidocaine injection and the longer-term analgesia provided. The lack of aversion in Day 0 sham calves is likely because they received half the lidocaine dose of Day 20 calves during conditioning. To conclude, we did not observe conditioned place preference 20 days after disbudding, but instead found that higher doses of lidocaine are aversive to uninjured animals. Despite this challenge with our paradigm, disbudded calves were willing to engage in the cost, providing indirect evidence that there is ongoing pain after 3 weeks.

### Analgesics on tap: efficacy assessed via self-administration and voluntary wheel-running in mice

*Jamie Ahloy Dallaire[1], Jerome T. Geronimo[1], Michael Gutierrez[1], Elin M. Weber[1,2], Cholawat Pacharinsak[1] and Joseph P. Garner[1,3]*
[1]Stanford University, Department of Comparative Medicine, 300 Pasteur Drive, Stanford, CA 94305-5342, USA, [2]Swedish University of Agricultural Sciences, Department of Animal Environment and Health, Box 234, 532 23 Skara, Sweden, [3]Stanford University, Department of Psychiatry and Behavioral Sciences, 401 Quarry Road, Stanford, CA 94305-5717, USA; jahloyda@stanford.edu

Large numbers of mice are used to test the efficacy of experimental drugs, but most drugs that seem to work in mice do not work in humans, often because outcomes are assessed differently. In the case of analgesics, the gold standard for efficacy is reduced self-reported pain severity, which correlates with drug self-administration using patient-controlled analgesia (PCA) machines. Mice could potentially provide "near-self-report" by displaying analogous behaviour: drug self-administration by injured (but not by non-injured) mice would support analgesic efficacy with high construct and predictive validity, better forecasting human outcomes and so reducing overall animal use. We developed a mouse version of PCA: a minimally-invasive, automated, 24-hour in-home-cage apparatus that monitors oral fluid self-administration. Drinkers in cage walls signal water or drug availability via light cues (as in human PCA) and track drinking by individual RFID-tagged mice. Mice (n=7 cages of 3) trained in three increasingly difficult phases learned to respond to individually-assigned light cues (by colour and position) while ignoring cues assigned to cagemates (Bonferroni-corrected post hoc tests after GLM: t≥2.90, $P \leq 0.004$). In three experiments (total n=20 cages of 2), mice with induced inflammation in one hind paw (Complete Freund's Adjuvant: CFA) consumed no more of the mildly unpalatable analgesic ibuprofen than their intact cagemates (Poisson GLZ: $\chi^2=2.48$, $P=0.115$). This apparent lack of efficacy conflicts with an alternative metric based on functional impairment, assessed in one of these experiments (n=6 cages): CFA mice wheel-ran less overall than controls, but became more likely to wheel-run as ibuprofen consumption increased (logistic GLZ: $\chi^2=4.15$, $P=0.042$). At sufficiently high dose, CFA mice ran as much as controls. Monitoring drinking by individual mice allowed us to use a powerful randomized block experimental design, housing mice of different treatments in the same cage. However, social transmission of taste preferences may hinder this approach: when both drinkers were available, mice were more likely to use the one their cagemate last used (logistic GLZ: $\chi^2=17.49$, $P<0.0001$), even controlling for all other relevant factors. This imitation is likely mediated by olfactory cues: the effect was stronger when drinkers contained water vs ibuprofen than water vs water ($\chi^2=5.21$, $P=0.023$). Mouse PCA may improve validity in analgesic drug development, but proof-of-concept will require a different combination of injury and drug. High-resolution automated monitoring allows assessment of efficacy by relating home-cage behaviour to drug intake over time. The apparatus, for which open-source schematics and software will soon be made available, can potentially be repurposed for other scientific and veterinary applications, including post-surgical analgesia and as a humane alternative to gavage.

**Efficacy of topical lidocaine on nursing behavior in castrated piglets**
*Mary Burkemper, Monique Pairis-Garcia and Steve Moeller*
*The Ohio State University, Animal Sciences, 2029 Fyffe Road, Columbus, OH 43210, USA;*
*burkemper.2@osu.edu*

Surgical castration is a procedure performed on nearly all male piglets destined for slaughter in the US swine industry as a means to eliminate unwanted breeding, decrease aggressive behavior and improve meat quality. Previous literature has consistently demonstrated that surgical castration is painful based on observed physiological and behavioral deviations of the piglet which presents welfare concerns. Therefore, the objective of this study was to evaluate the effect of administering two analgesic agents (oral meloxicam and topical lidocaine spray) on suckling behaviors in piglets undergoing surgical castration. Two hundred eighty nine male piglets across 42 litters were castrated between 3 and 7 days of age. Piglets received one of four treatments immediately following surgical castration: (1) No pain mitigation (C; n=74); (2) NSAID only, oral meloxicam, 1.0 mg/kg PO, (M; n=72); (3) topical anesthetic of lidocaine spray (L; n=71); or (4) combination of oral meloxicam and lidocaine spray (X; n=72). Within a litter, at minimum, each treatment was assigned to a piglet based on the four or eight pigs closest in weight. Remaining male pigs (5th, 6th, 7th, or 9th plus) within a litter were allocated to treatment across litters, matching weight distribution across treatments to the extent possible. Behavior was recorded by live observation of each individual piglet using five-minute scan samples for a five-hour period on the day of castration and for two consecutive days following, (i.e. piglets were observed for hours 0-4, 24-28, and 48-52 post-castration). Piglet behaviors were mutually exclusive and grouped into four categories; active, inactive, suckling, and pain behavior (tail wagging, rump scratching, trembling, or prostration). No significant differences were found in active, inactive, or pain behaviors across treatments during the entire observation period. However, piglets in the lidocaine treatment suckled less than all other treatment groups over the entire observation period ($P<0.05$), with a 17.3% probability of suckling at any given scan compared to 18.9, 19.5, and 19.0% (C, M, X) respectively. In this study, topical anesthetic in the form of lidocaine spray applied to the wound immediately following surgical castration resulted in a decrease in the probability of piglets suckling, possibly due to the large percentage of alcohol in the spray. This suggests that the utilization of a lidocaine spray does not mitigate deviations to the piglet's behavioral repertoire of suckling.

## The effect of rearing environment on laying hen keel bone impacts sustained in enriched colony cages

*Allison N. Pullin[1], Mieko Temple[2], Darin C. Bennett[2], Richard A. Blatchford[1] and Maja M. Makagon[1]*
*[1]University of California, Davis, Animal Science, 1 Shields Ave, Davis, CA 95616, USA, [2]California Polytechnic State University, Animal Science, 1 Grand Ave, San Luis Obispo, CA 93407, USA; apullin@ucdavis.edu*

Laying hens housed in enriched colony cages sustain impacts to their keel bone from collisions with objects and other birds, which can result in keel bone fractures. Aviary (A) rearing environments, which offer greater spatial complexity as compared to rearing cages (C), have been associated with fewer keel bone fractures in hens that are later housed in enriched colony cages. The objective of this study was to evaluate whether this reduction in keel bone fracture incidences reflects differences in the way that C and A birds interact with their later environments. Lohmann LSL-Lite hens were reared in either C or A until 19 weeks of age and then moved into enriched colony cages (averaged 24 birds/cage), each containing two elevated perches of different heights (n=6 cages/treatment). At 20 weeks of age, keel-mounted tri-axial accelerometers were fitted onto three hens per cage for 11 days to record impacts sustained at the keel bone. Video recorded continuously over the 11-day period was used to pair behavior with corresponding impact incidences. A total of 425 identifiable impacts to the keel (C: 288, A: 137 impacts) were categorized as occurring during grooming, aggression, mass-scattering events, or collisions with other birds or objects in the environment. The percentage of impacts within each behavioral category was calculated for each hen. Data were checked for normality, log transformed if needed, and analyzed using a two-sample t-test with PROC TTEST in SAS 9.4. Rearing environment did not affect the percentage of keel impacts identified as grooming (C: 27.3±6.3, A: 16.7±6.4%, mean ± S.E.; $P>0.05$), aggression (C: 12.0±2.1, A: 17.9±4.2%; $P>0.05$), or mass-scattering (C: 13.1±3.7, A: 8.0±3.3%; $P>0.05$). More impacts were associated with collisions for A than C hens (C: 38.8±3.3, A: 55.2±4.0%; $P=0.01$). The objects that birds collided with (i.e. low perch, high perch, bird, or floor) did not differ by rearing type; however, the actions causing collisions did. More of the collisions experienced by A vs C hens were caused by unsuccessful ascents from the low perch to the high perch (C: 0.02±0.001, A: 0.8±0.08%, mean ± 95% CI; $P=0.02$). Running into objects or conspecifics was associated with more of the collisions experienced by C than A hens (C: 20.4±9.5, A: 0.4±0.03%, mean ± 95% CI; $P=0.01$). It is not clear from these results whether these differences are due to differences in spatial navigation abilities or resource use by hens from the two rearing treatments. One possible explanation is that hens reared in the more complex aviary environment may attempt navigating vertical resources more than cage-reared birds. Our future work will evaluate this possibility and assess the long-term effects of rearing environment on hen behavior through the end of lay.

## Ramps to facilitate access to different tiers in aviaries reduce keel bone and foot pad disorders in laying hens

*Frank Tuyttens[1,2], Nikki Mackie[3], Sofie De Knibber[1], Evelyne Delezie[2], Bart Ampe[2], Bas Rodenburg[4] and Jasper Heerkens[2,5]*
[1]*Ghent University, Faculty of Veterinary Medicine, Salisburylaan 133, 9820 Merelbeke, Belgium,* [2]*Flanders Research Institute for Agriculture, Fisheries and Food (ILVO), Animal Sciences Unit, Scheldeweg 68, 9090 Melle, Belgium,* [3]*University of Bristol, School of Veterinary Science, Langford, BS40 5DU, United Kingdom,* [4]*Wageningen University, Behavioural Ecology Group, P.O. Box 338, 6700 AH Wageningen, the Netherlands,* [5]*Aeres University of Applied Sciences, De Drieslag 4, 8251 JZ Dronten, the Netherlands; frank.tuyttens@ilvo.vlaanderen.be*

Keel bone and foot pad disorders are major welfare problems in laying hens. The prevalence of both disorders tend to be highest in housing systems that are commonly perceived to offer improved animal welfare, such as aviary systems. We tested whether providing ramps, that allow hens to negotiate height differences without flying or jumping, reduces the prevalence of these disorders, first in an experimental setting and subsequently in an on-farm study. The experimental study was a 2×2 design in which 16 pens were equipped either with or without ramps between perches and nest boxes (8 pens/treatment) and housed with either 25 ISA Brown or Dekalb White birds per pen (in total 200 birds/hybrid). In the on-farm study, 18 commercial aviaries (7 with ramps and 11 without ramps) were visited when the hens were 42-85 weeks old. Besides flock level behavioural observations of up- and downwards movements, all birds of the experimental setting and a sample of ca. 70 hens per flock in the on-farm study were scored for keel bone fractures & deformations and for foot health. Data were analysed using glmer or lmer models in R with ramp status, hybrid and age, and their two-way interactions, as fixed factors (if significant) and with flock as random factor. Although the behavioral observations revealed no clear differences, in both the experimental and on-farm study, the prevalence of keel bone fractures ($P<0.001$, $P=0.031$) and bumble foot (in both studies $P<0.001$) were lower in flocks with ramps compared to flocks without ramps. In the experimental setting ramps also reduced foot pad hyperkeratosis ($P=0.001$) and dermatitis ($P<0.001$). Assuming no expectation bias due to unavoidable observer knowledge of treatments, these results show that ramps reduce, but do not eliminate, keel bone and foot pad disorders in laying hens housed in aviary systems. This suggests that at least part of these disorders relate to the difficulty hens experience when negotiating height differences in such complex housing systems when no ramps are provided.

## Startle reflex as a welfare indicator in laying hens

*Misha Ross, Alexandra Harlander and Georgia Mason*
*University of Guelph, Department of Animal Biosciences, 50 Stone Rd E, Guelph, ON N1G 2W1,*
*Canada; mross19@uoguelph.ca*

Startle reflexes are rapid, involuntary responses to intense stimuli with rapid rise times (e.g. sudden sounds). In mammals, they track affective states (increasing in magnitude in negative states, and decreasing in positive ones), but whether these findings generalize to any avian species was unknown. We therefore investigated the startle reflex as a potential indicator of affective state in laying hens. We hypothesized that adult laying hens in positive affective states would show diminished startle magnitudes. We manipulated affective state in two ways: first, by training hens to associate the startle test apparatus with reward; and second, by allowing hens long-term access to preferred housing conditions. In Experiment 1, 24 ISA Brown laying hens were individually placed in a dimly lit chamber and exposed to a xenon flash, during which their baseline startle magnitudes were measured using a force plate. 16 hens then received 20 sessions of reward conditioning in the dim chamber (meal worms and sweet corn), while the remaining 8 were used as controls. To assess the effectiveness of reward conditioning, latencies to approach the chamber were measured. The startle test procedure was then repeated for all hens. Conditioned hens' approach latencies were shorter than control hens' ($F_{1,23}$=42.02, $P$<0.0001), confirming that they formed a positive association with the test chamber. However, contrary to our prediction, reward conditioning did not reduce hens' startle magnitudes compared to controls (11.08±11.29 vs 10.57±11.80 N [LSMs ± SEs]; $F_{1,17}$=0.15, $P$=0.702). In Experiment 2, 96 ISA Brown laying hens were randomly divided into 24 groups. 12 control groups were housed in 1.5 square meter pens with a nest box, a perch, bedded with wood shavings. 12 enriched groups were kept in 10 square meter enclosures with a priori preferred resources including: feeding enrichments, dust bathing substrates and perches of various heights and diameters. Enriched hens also had free access to a proxy control pen, the use of which was assessed using scan sampling. After 5 weeks in their respective housing conditions, hens were individually startled using the same procedures as Experiment 1. Groups of enriched hens spent little time in their proxy control pens even after correcting for its smaller floor area ($Z_{1,11}$=-39.0, $P$<0.001), confirming that the resources were preferred and thus produced more positive affect. As predicted, the startle magnitudes of enriched hens were significantly lower than control hens (1.79±1.56 vs 19.65±1.55 N; $F_{1,23.53}$=104.00, $P$<0.0001). Experiment 2's results thus support the validity of the startle reflex as a hen welfare indicator: the first ever demonstration that findings in mammals also hold for birds. However, the lack of a similar pattern in Experiment 1 indicates that more research is needed to assess and improve this indicator's sensitivity. Ethical approval was granted under Animal Use Protocol #3763.

## Human-hen relationship on grower and laying hen farms in Austria and associations with hen welfare

*Susanne Waiblinger, Katrina Zaludik, Jasmin Raubek, Bettina Gruber and Knut Niebuhr*
*University of Veterinary Medicine Vienna, Institute of Animal Husbandry and Animal Welfare,*
*Department/Clinics for Farm Animals and Veterinary Public Health, Veterinärplatz 1, 1210*
*Vienna, Austria; susanne.waiblinger@vetmeduni.ac.at*

In several farm animal species sequential relationships between stockperson attitude, behaviour and animal behaviour and welfare have been shown; but this was lacking for laying hens in non-cage systems. The aim of this study was to investigate, both in rearing and laying hen units, the variability in farmer's attitudes, handling practices, the hen-human relationship and their associations as well as with measures of welfare. 25 grower flocks (16-17 weeks of age) and 50 layer flocks (30-40 weeks of age) were visited once. The hens' relationship towards humans was assessed by measuring their reactions to humans in standardized, validated tests (avoidance distance (AD) test, TouchTest (TT)). Information on work organisation, handling practices and attitudes of the main stockperson was collected by a questionnaire, farm/flock characteristics and management practices via a pre-structured interview. Plumage condition, skin injuries and housing characteristics were evaluated directly, and mortality from farm records. Attitude factors were derived from principal component analysis on single items. There was considerable variation in hens' reactions towards humans as well as in stockpeople's attitudes and behaviour towards the hens. The average number of animals touched per trial ranged from 0 to 3 in TT (median growers/layers:1.5/0.9) and AD ranged from 0 to 1.5 m (layers: 0.5 m, not tested in growers), thus covering the full possible ranges. Most stockpeople of grower flocks rated quite high in positive general attitude (med, min-max: 5.7, 3.7-7; Likert scale on single items: 1, completely disagree; 7; completely agree), as did stockpeople of layer flocks (4.8, 1-6.7); however there was also agreement with negative general attitudes (4, 2-5.67). Most stockpeople agreed that contact with hens is important (ImportantContact) as well as on the importance of regular contact, intensive care and housing (ImportantCare) for hen behaviour. Working routines, contact with hens and handling practices correlated with attitudes, the hen-human relationship and welfare. E.g. stockpeople of layer flocks that declared that they caught their hens regularly ranked higher in ImportantCare (Mann-Whitney U, $P=0.045$), and in these flocks more hens could be touched in TT ($P=0.010$) and more hens had an AD of 0 ($P=0.033$). The percentage of hens with feather damage correlated positively with the median of AD ($r_s=0.43$, $P=0.002$) and negatively with hens touched in TT ($r_s=-0.31$, $P=0.028$) and, in growers, with agreement on ImportantContact ($r_s=-0.43$, $P=0.040$). In laying hens, negative general attitude correlated positively to the percentage of hens with feather damage ($r_s=0.37$, $P=0.008$) and mortality ($r_s=0.31$, $P=0.047$). Our results confirm sequential relationships found in other species and suggest that human attitudes and subsequent behaviour are important for an improved hen-human relationship and hen welfare in non-caged hens.

### Laying hen preference for consuming feed containing excreta from other hens

*Caroline Graefin Von Waldburg-Zeil, Nienke Van Staaveren and Alexandra Harlander-Matauschek*

*University of Guelph, Department of Animal Biosciences, Ontario Agricultural College, 50 Stone Road E, Guelph, Ontario N1G 2W1, Canada; cgraefin@uoguelph.ca*

Increasing animal welfare concerns have led to changes in laying hen housing, including the design of litter-based systems to allow foraging behaviour. Foraging is an activity in which a series of behavioural acts lead to the consumption of feed. Foraging substrate becomes increasingly soiled throughout a laying hen's life, increasing the likelihood that hens will consume portions of excreta found in the foraging substrate. However, no study has investigated the relative preference of laying hens for consuming feed that has been mixed with excreta. Therefore, it was our goal to investigate whether laying hens would consume feed mixed with increasing percentages of excreta. Forty-eight 71-wk old White Leghorn laying hens (LSL strain, n=24; UCD-003 strain, n=24) were transferred from group housing on a combination of deep litter (pine shaving-excreta mix) and slatted floors, to individual housing in enriched cages, and given daily access to standard commercial feed (18% CP, 4.2% Ca, 0.44% P, 0.18% Na, 2,900 kcal/kg ME, 0.89% lysine, 0.38% methionine) mixed with increasing percentages of fresh hen excreta (0, 33, 66 and 100% excreta diets). Fresh excreta were collected daily from non-experimental adult laying hens receiving the same feed and kept under the same housing conditions. Diets were placed in four equally sized containers (134.3±10.16 g diet/container) and containers were simultaneously presented in a systematically varied order in front of the cage. Hens were habituated to the experimental set-up for one week, after which the amount of substrate consumed from each diet was recorded for 3 weeks to determine the relative preference for each diet. Generalized linear mixed models (PROC GLIMMIX, SAS V9.4) were used to determine the effect of diet, strain and their interaction on the percentage of substrate consumed, with bird identification included as the repeated subject. Both LSL and UCD-003 hens showed a clear decrease in the amount of substrate ($P<0.001$) consumed as the percentage of excreta increased. The majority of the substrate in the 0% excreta diet was consumed (69±2.64%), followed by the 33% excreta diet (28±3.53%) and finally diets containing 66% excreta (10±1.01%) and 100% excreta (8±0.32%). Nevertheless, birds consumed on average 61.3 g of diets containing excreta per day. This study demonstrated a clear preference in laying hens for consuming substrate without excreta. However, considering the amount of excreta consumed in diets, further studies are needed to understand the causes and consequences of excreta consumption and its implication for the management of foraging substrates in laying hens.

## The reliability of palpation, x-ray, and ultrasound techniques for the detection of keel bone damage

*Linnea Tracy[1], Mieko Temple[2], Darin Bennett[2], Kim Sprayberry[2], Maja Makagon[3] and Richard Blatchford[3]*
*[1]University of Pennsylvania, School of Veterinary Medicine, Philadelphia, PA 19104, USA, [2]California Polytechnic State University, San Luis Obispo, Animal Science, 1 Grand Ave, San Luis Obispo, CA 93407, USA, [3]University of California, Davis, Animal Science, Center for Animal Welfare, One Shields Ave, Davis, CA 95616, USA; rablatchford@ucdavis.edu*

As the egg industry transitions from housing laying hens in cages to cage-free production systems, the incidence of keel bone damage has increased. Measuring keel bone damage on live birds is generally achieved via palpating the keel bone. However, this method is problematic due to its subjective nature and intensive training it requires. The objective of this study was to investigate the inter- and intra-observer reliability of assessors trained with feedback of accuracy, as well as the accuracy of portable radiographic and sonographic examination. At the time of depopulation, 50 103-week old Lohmann LSL-Lite hens were euthanized and immediately palpated by 4 assessors. The keel bones of the hens were then sonographed, radiographed, palpated a second time, and finally, keel bones were dissected and cleaned for visual examination. For each method, keel bones were scored for the presence of deviations, fractures, and tip fractures of the keel bone. Intra-rater reliability was tested using a Cronbach's Alpha test while inter-rater reliability was tested using a Fleiss' Kappa test for each type of keel bone damage. Using the dissection scores, sensitivity, specificity, positive predictive value and negative predictive value were calculated for both the sonography and radiography scores. All statistical tests were performed using R 3.3.1. Inter-observer reliability in the first round of palpations was 0.39 for deviations, 0.53 for fractures, and 0.12 for tip fractures. Inter-observer reliability in the second round of palpations was 0.36 for deviations, 0.55 for fractures, and 0.12 for tip fractures. Intra-observer reliabilities using palpation ranged from 0.58 to 0.79 for deviations, 0.66-0.90 for detecting fractures, and 0.37 to 0.87 for tip fractures. Inter-observer reliabilities using palpation were 0.39 for deviations, 0.53 for fractures, and 0.12 for tip fractures. Sensitivity of sonography and radiography were generally high for fractures (75.0, 85.7%) and tip fractures (90.9, 84.4%), but low for deviations (50.0, 60.9%). Specificity of sonography and radiography was generally high across deviations (75, 72.7%), fractures (78.6, 81.5%), and tip fractures (67, 100%). Intra- and inter-reliability were low and highly variable even when assessors received information on the accuracy of their scoring. This suggests palpation is likely not a good method of detecting keel bone damage. Both portable sonography and radiography show promise as methods of detecting fractures, but not deviations. However, the effects of restraining a live bird for these methods needs to be examined.

**Focal hen behaviour in a free-range commercial setting, does the presence of a keel fracture affect standing, walking?**

*Francesca Booth[1], Stephanie Buijs[1,2], Gemma Richards[1], Christine Nicol[3] and John Tarlton[1]*
*[1]University of Bristol, Langford House, Langford, BS40 5DU, United Kingdom, [2]Agri-Food and Biosciences Institute (AFBI), Large Park, Hillsborough, BT26 6DR, Ireland, [3]Royal Veterinary College, Hawkshead Lane, Hatfield, AL9 7TA, United Kingdom; francesca.booth@bristol.ac.uk*

Free ranging facilitates behavioural and physiological stimulation and allows freedom to choose between internal and external environments. However, increased activity, especially increased flights, and greater exposure to physical hazards, can lead to injurious collisions of the keel bone. Up to 80% of hens in free range systems in the UK suffer fractures by end of lay (72 weeks), repair may take weeks or months, and during this time the hen could experience pain and reduced mobility, potentially reducing access to feed and water. During the study 81 focal British Blacktail hens (flock size 2,000) from 6 commercial flocks (approximate total number of hens 12,000) were tagged and behaviour observed and recorded for each at 21, 27, 33, 37 and 42 weeks of age. Within each flock, 7 hens were randomly selected inside the house and 7 more outside on the range (3 hens in total were later excluded due to mortality). Hens were housed in 6 identical flat deck mobile systems with a 2-hectare range attached to each. Hens were assessed for the presence or absence of a keel fracture, by palpation, before (20, 26, 32, 36 and 40 weeks of age) and after observation periods (same weeks as observations). Continuous focal sampling (5-min/hen/week) was used to record standing (inside/outside), walking (inside/outside) and foraging (inside/outside) on a handheld device. Observations were divided over the entire day (which varied with season), rotating hens over the different parts of the day. At the end of the observation period (43 weeks) focal birds were euthanised and the keel excised for keel fracture presence and severity scoring (a five-point scale of 0-4, where 0 is no damage, and 4 is severe damage). Out of the 81 focal hens, 47 had a keel fracture (score between 1 and 5) and 34 did not have a keel fracture (score 0). Standing, walking and foraging behaviours were root transformed and analysed in a mixed model with fracture status, age and their interaction as fixed factors and hen as a random factor. Hens with a keel fracture (score 1-5) at post-mortem (43 weeks of age) walked less during observations (21-42 weeks) than those without a fracture [$P=0.02$, LSMEANS ± SEM: 16±1 vs 22±2% of total time]. No effect of fractures on standing [$P=0.9$] or foraging [$P=0.8$] was identified. Age effects on behaviours were not linear. Standing occurred significantly more in week 21 [$P\leq0.03$, 22±3%] and week 42 [$P\leq0.03$:18±2%] than in week 33 [$P\leq0.03$, 10±2%]. Foraging increased after 21 weeks of age [$P\leq0.02$, 6%±2] and decreased at 42 weeks of age [$P\leq0.02$, 19%±3]. No effect of age on the time spent walking was found [$P=0.6$]. The results show that keel fracture status at 43 weeks is associated with duration of walking behaviour between 21 and 42 weeks of age, future work will be necessary to determine if keel fractures led to reduced walking (e.g. because of pain) or if less walking led to an increased fracture incidence (as walking stimulates bone strength and skeletal resilience).

**Laying hens that differ in range use behave similarly outside, but differently inside**

*Stephanie Buijs[1,2], Francesca Booth[2], Gemma Richards[2], Christine J. Nicol[3] and John F. Tarlton[2]*
[1]*Agri-Food and Biosciences Institute (AFBI), Large Park, BT26 6DR Hillsborough, United Kingdom,* [2]*University of Bristol, Langford House, BS40 5DU Bristol, United Kingdom,* [3]*Royal Veterinary College, Hawkshead Lane, AL9 7TA Hatfield, United Kingdom; stephanie.buijs@afbini.gov.uk*

Outdoor access allows expression of behaviours that can be performed less satisfactorily indoors. However, time spent outside varies greatly and it is unknown if hens that range less go out exclusively to perform specific behaviours. If so, adapting the range to facilitate such behaviours could stimulate range use, which is often low. We observed 84 hens in 6 commercial flocks of 2,000 hens each. Three hens died and were excluded. All other hens were categorized as 'high rangers' (n=23), 'medium rangers' (n=37) or 'low rangers' (n=21) based on 4 separate weeks of automated range use monitoring using hen-mounted light sensors. A hen was classified as a high ranger if she was within the top 5 of range users amongst the 14 monitored hens in her flock, in at least 3 out of 4 weeks. Low rangers were selected similarly from the bottom 5. All others were classified as medium rangers. As range use rankings were more consistent between weeks in some flocks than others, and inconsistent hens ended up in the medium category, the number of high, low and medium rangers differed slightly between flocks. During direct focal observations at 21, 27, 33, 37 and 42 weeks of age, behaviours and locations (in/out) were recorded (5 min/hen/age) and averaged per hen. Hens were identified using coloured leg rings and a locator device (TileMate). Data were analysed by Kruskal-Wallis tests. Where differences between high and low rangers are indicated, medium rangers had intermediate values. High rangers spent more of their total time foraging outside than low rangers (median (IQR): 44% (29-50) vs 16% (10-30), $P<0.001$) and indoor foraging was negligible. Ranging style did not affect total walking time, but high rangers walked more outside and less inside than low rangers (20% (10-24) vs 8% (9-15) and 4% (1-7) vs 13% (9-15), both $P<0.01$). Crucially, ranging style did not affect how outdoor time was spent ($P\geq0.3$, foraging 54% (37-67), walking 26% (16-32), standing 13% (6-23), preening 0% (0-2)). This may suggest that, rather than high and low rangers using the outdoor area for different behaviours, they differ in their motivation to perform the same behaviours, or in how they balance this motivation with other factors. Ranging style affected indoor behaviour: when inside, high rangers tended ($P=0.09$) to spend less time eating/drinking than low rangers. Combined with less time spent inside overall, this meant they spent far less of their total time eating/drinking than low rangers (1% (0-13) vs 19% (12-23), $P<0.001$). Ranging style did not affect post-experimental bodyweight ($P=0.8$), suggesting that high rangers were more efficient eaters/drinkers, ate more after dark (when un-observed), acquired nutrients from the range, or started out heavier (no start weights were acquired). Increasing food availability on the range, or limiting food availability inside, may be key in stimulating range use in hens that otherwise range little, although potential trade-offs (e.g. lowered biosecurity or feed intake) also need to be explored.

**The effects of body weight on the diurnal range use and latency to use the range in free range laying hens**

*Terence Sibanda, Manisha Kolakshyapati, Johan Boshoff and Isabelle Ruhnke*
*University of New England, Environment and Rural Science, 60 Madgwick Rd, 2350, Armidale,*
*Australia; tsibanda@myune.edu.au*

Understanding range usage of free-range laying hens is essential to address the needs of the hens and apply feeding strategies most efficiently. While heavier hens might be of increased social status and therefore dominate range access after pop hole opening, lighter hens may still access the range at different times of the day. The objective of this study was to determine the impact of body weight on diurnal range usage and the latency to range. A total of 3,125 Lohmann Brown hens were randomly selected at the age of 16 weeks and placed in 5 partitioned subsections with equal number of birds (5 replicates) within a commercial free range shed holding a total of 40,000 hens. All hens were individually weighed, and marked using a numbered leg band with an individual radio frequency identification (RFID) microchip. Range access was given to all hens from 9 am until 5 pm daily. Pop holes covered the entire length of the shed, hence hens of the partitioned subsections had 3.6 m pop hole area available. The range was structured using artificial shade cloth covering 5% of the ranging area. A total range area of 2.1 ha allowed for a stocking density of 1,500 hens/ha. Within each replicate, hens were grouped based on their body weight in either group 1 (hens below the average body weight; mean ± SEM: 1.28±0.002 kg), or hens of group 2 (heavier than the average body weight; 1.43±0.001 kg). The final hen numbers used for statistical analysis excluding hens with lost or default RFID tags were n=1,450 and n=1,451 for group 1 and 2, respectively. The average values per replicate were used for statistical analysis. Individual range access was monitored for 7 consecutive days when hens were 17 weeks of age. Linear regression models were used examine the effect of body weight on the percentage of hens that accessed the range per hour, frequency of range visits, and latency to range using SPSS Statistics 24v. Significance was set at $P<0.05$. Significantly more hens of group 2 were present on the range at 9 am, 10 am, 11 am, 12 pm, 1 pm, 2 pm, 3 pm, 4 pm, and 5 pm, compared to group 1 ($P<0.05$). Group 2 hens visited the range significantly more frequently 14.6±0.31 visits per hen/day compared to 13.6±0.36 visits per hen/day for lighter hens ($P=0.038$). There was no significant difference in latency to range between hens of group 1 and hens of group 2 ($P=0.248$). In conclusion, range usage can be significantly influenced by hen body weight at time of placement.

## Feeding black soldier fly larvae to laying hens: effect on pecking and foraging behaviour

*Marko A.W. Ruis[1], Jasper L.T. Heerkens[2], Laura Star[2], Henry J. Kuipers[1], Femke Kromhout[1] and Janmar Katoele[3]*
[1]*VHL University of Applied Sciences, Department of Animal Management, Agora 1, 8934 CJ Leeuwarden, the Netherlands,* [2]*Aeres University of Applied Sciences, De Drieslag 4, 8251 JZ Dronten, the Netherlands,* [3]*Wadudu Insect Centre, Noordveen 1, 9412 AG Beilen, the Netherlands; marko.ruis@wur.nl*

Inclusion of insects in animal feed is promising for poultry behaviour and welfare. The aim of this study was to feed black soldier fly (BSF; *Hermetia illucens*) larvae to Bovans Brown laying hens and to study the effects on pecking and foraging behaviours. Live BSF larvae were fed to hens from 20-36 wks of age. Four feeding treatments were used, with 20 pens having 20 hens each (n=5/treatment). Space per pen was 3 m$^2$: 1/3 litter of wood shavings, 2/3 raised wooden slats. Unlimited water and mash feed were provided. Adult larvae were either provided daily in the feed bin (24 g; T1), spread daily in the litter (24 g; T2), not being fed (T3), or crawled out through the slats after development in the manure (T4). For this purpose, 4-5 days old larvae (about 70 g) were spread in the manure, twice a wk. Feather and resource pecking behaviours were observed at 19, 25, 30 and 35 wks of age. On 2 days per observation wk, the behaviour was noted 16 times in total between 10:00-14:30 h, through 1-min continuous samplings (one/zero sampling). At the end of the 1-min observations, numbers of hens engaged in specific other behaviours were scored. Plumage and skin condition were evaluated at 19, 28 and 36 wks of age, as a reflection of feather pecking. The back/neck, tail and cloaca area were checked, with scores of 0-1-2 (range: no damage (0) to larger damage (2); feathers: bald spot >5 cm; skin: wound >2 cm). Data were analysed using SPSS 23. GLMMs were used, with treatment, age and their interaction as predictors (random intercept fitted for pen number). No differences were observed in feather pecking, but damage by feather pecking was decreased in T4 hens: they maintained the highest quality of tail feathers (score 0; treatment × age interaction: F(6, 48)=2.437; P=0.039). At 36 wks of age, scores of 0 were more often seen in T4 birds (39.3±16.8%) than in T3 birds (1.8±2.0%; P<0.05). At this age, T4 hens also had a lower frequency (P<0.05) of severe damage (score 2) to cloaca feathers (13.0±19.4%), compared to T2 (83.7±23.1%) and T3 hens (89.1±16.3%) (treatment × age interaction: F(3,32)=6.622; P<0.001). Treatment affected pecking to the slatted floor (F(3, 941)=11.408; P<0.001): T4 hens generally showed more pecking (all comparisons P<0.004) to the slatted floor (2.0±0.5%) than the other hens (T1: 0.2±0.1%; T2: 0.0±0.4%; T3:0.2±0.1%). For pooled active behaviours, the effect of treatment depended on age of the hens (interaction: F(9, 1257)=2.430; P=0.010). T4 hens showed the highest activity (P<0.05), compared to T2 (at age 25 wks: 61.9±2.1 vs 55.2±2.1%) and T1 and T3 hens (at age 30 wks: 76.7±1.7 vs 69.6±2.0% (T1) /67.4±2.0% (T3)). Results suggest that live BSF larvae are a valuable source of environmental enrichment. Feeding insect larvae seems promising for poultry welfare with prolonged time of foraging and distraction, as in T4 hens where activity and foraging behaviour increased, and damage from injurious pecking behaviour decreased.

**The probiotic *Bacillus subtilis* does not improve laying hen performance in a modified puzzle box cognition test**

*Jacquelyn Jacobs and Heng Wei Cheng*
*USDA ARS Livestock Behavior Research Unit, West Lafayette, IN 47907, USA; jacob115@purdue.edu*

In the U.S., numerous laying hen producers have committed to providing cage-free systems within the next few years, introducing a more complex environment to the birds. Cognitive ability, including reasoning, learning, and memory, will be particularly important to ensure effective resource use. Probiotics are beneficial bacteria that improve the microbial environment in the intestine and have various purported benefits on the host through the microbiota-gut-brain axis. Several studies have suggested that probiotics can improve cognitive abilities in mice, but this has not yet been tested in laying hens. The objective of this pilot study was to determine if the probiotic *Bacillus subtilis* would increase performance of laying hens in a modified puzzle box test, validated to assess reasoning, learning and memory skills in mice. At one day of age, birds were assigned to groups of 5 and housed in conventional cages (439 cm$^2$) with feed and water provided *ad libitum*. Cages were randomly assigned to either a control diet (n=28), or a diet supplemented with 400 ppm *B. subtilis* (n=28). The cognition test commenced when birds were 34 wks of age. In brief, birds were individually trained to walk the length of a 2×1.2 m pen. The 'goal zone' was rewarded with a bowl containing live mealworms (mw). Latency to reach the goal zone and consume the first mw was recorded during all training sessions. Training concluded when the bird completed three successive walks to the goal zone in six seconds or less. The test pen arrangement was identical to training, except that the bowl was covered tightly with paper towel and access to the mw could only be achieved by pecking or clawing the paper towel. To reduce novelty, a paper towel was fixed tightly to one side of the birds' home cage 24 hrs prior to training. The test consisted of three identical trials (T1-T3); T1 began immediately after training (reasoning), T2 immediately after T1 (learning), and T3 occurred 24 hrs after T2 (memory). Latency to consume the first mw at each trial was recorded. A Fishers exact test (SAS 9.4) was used to determine if the number of required training sessions differed between treatment groups. A linear model was developed to determine the association between treatment and various performance parameters. Overall, birds performed faster on T2 (19.45 s±1.18) and T3 (15.61±1.18 s) compared to T1 (31.72±2.22 s; $P<0.001$), with no difference in performance between T2 and T3 ($P=0.45$). The number of required training sessions and overall performance ability did not differ between treatment groups ($P=0.54$; $P=0.84$, respectively). This test procedure has potential for assessing a simple form of learning and memory in laying hens, although it requires validation for further use in poultry. The findings of this study suggest that the probiotic *B. subtilis* does not improve learning and memory ability, however, it is possible that an alternative probiotic may have a positive effect on these aspects of cognition considering the variety of probiotics available and their various effects on the microbiota-gut-brain axis.

## The unity and differentiation in mammalian play

*Marek Špinka*
*Institute of Animal Science, Department of Ethology, Pratelstvi 815, 104 00 Prague, Czech Republic; spinka.marek@vuzv.cz*

Play behaviour is almost universal among mammalian species, in contrast to other vertebrate classes. Furthermore, despite large variation in its forms and functions, mammalian play is clearly distinct from all other 'serious' types of behaviour. It follows that play behaviour originated very early in mammalian phylogeny and must have a specific and powerful adaptive potential that kept it firmly present in the gene pool and expressed in the behavioural phenotype of mammals throughout their history. Here I posit that this adaptive potential of play resides in the fact that its putative original form, namely the juvenile locomotor-rotational play, provides numerous opportunities to develop derived forms of play that fulfil diverse functions in various species of mammals. In order to highlight the adaptive potential of juvenile mammalian play, I will briefly review its specific features. First, juvenile mammalian play contains many distorted movements and self-handicapping actions that result in temporary loss of control over own body kinematics. Second, a specific proximate affective mechanism ('fun') motivates and rewards juvenile play. Third, juvenile mammalian play is highly repetitive, i.e. once a play element or sequence is performed, a strong tendency is triggered to immediately repeat it over and over. The three aspects are linked together. Loss of control, which would in 'serious' context be aversive and block the repetition of the behaviour, does the opposite in play: it contributes to the fun affect and thus promotes the repetition of current play elements. The combination of these three aspects results in juvenile mammalian play being performed in the form of intense, enjoyable and repetitive variations on a particular behavioural theme. I propose that the original function of juvenile mammalian play was (and in many extant species still is) to pre-map and train own locomotive skills vis-a-vis the unpredictable nature of the future interactions with the environment. I will illustrate the diversification of play from the original form with three scenarios: the object play, where the play motivation gets merged with exploration; the social play, where contagiousness of play can lead to all degrees of play sociality from mere time coordination to highly elaborate communicative interactions; and mental play that includes cases such as teasing and, in humans in the very least, humour and pretend play. I will finish the presentation with stressing that despite the common phylogenetic root, the concrete forms and roles of play are vastly diverse in particular mammalian species. Consequently the role of play behaviour as an indicator and/or promoter of animal welfare varies across species of domestic and captive mammals, ranging from almost negligible to crucially important. Therefore, the between-species differences in play are as an important topic for focused research as is the common origin.

## Do any forms of play indicate positive welfare?

*Georgia J. Mason[1], Julia Espinosa[2] and Jamie Ahloy Dallaire[3]*
*[1]University of Guelph, Animal Biosciences, Guelph, Ontario, N1G 2W1, Canada, [2]University of Toronto, Psychology, 100 St. George Street, Ontario, M5S 3G3, Canada, [3]Stanford University, Comparative Medicine, Stanford, California, 94305-5342, USA; gmason@uoguelph.ca*

To ensure good animal well-being, we need objective welfare indicators that are sensitive to positive affective states. Play is often held up as such an indicator. But are any forms of play really useful, valid tools for revealing positive welfare? We argue that hardly any research supports this idea, primarily because to date, the right data have not been collected. This gap in evidence is caused by two major challenges that face researchers wanting to test the validity of play as an indicator of positive affect. The first is that identifying conditions that induce absolutely positive states in animals (rather than merely relatively positive states) is surprisingly difficult. Studies comparing enriched and barren conditions, for example, may be revealing more about the effects of reducing boredom and frustration, than about any increase in positive affect. To induce unambiguously positive states, we therefore need clear, objective ways to operationalize 'pleasure', 'happiness' or 'contentment' in animals; and furthermore, we then need to be able to experimentally induce differing degrees of positive affect to assess empirically whether these influence play. To date, the few studies to have plausibly done this suggest that two forms of play are sensitive to positive affective states: social (rough-and-tumble) play in rats and locomotor play in piglets. But most welfare-oriented play research has only compared animals in sub-optimal states with animals in better ones, and the same holds for research on the effects of affective states on human play (which typically compares ill or distressed children with controls). The second research challenge is that the term 'play' covers a heterogeneous group of behaviours, such that data from one form or species cannot necessarily be directly applied to another form or species. This helps explain why, despite the rat and piglet examples above, some forms of play increase rather than decrease in animals and humans in aversive situations (examples including object play in food-deprived kittens, and re-enactment play in traumatised children). The welfare significance of each type of play must therefore be validated *de novo* on its own merits. Nevertheless, data from children highlights some welfare-relevant qualitative aspects of human play that so far have been over-looked in animal play research (e.g. the degrees to which play behaviour is 'fragmented', oscillating between non-play and play activities). These might provide metrics that overcome the heterogeneity problem. Overall, we hope our suggestions will pave the way for more nuanced validatory research into the affective correlates of play, which might then identify qualitative or quantitative aspects of sub-types of animal play that do indeed indicate positive affective states.

## Determining an efficient and effective rat tickling dosage

*Megan R. Lafollette[1], Marguerite E. O'Haire[2], Sylvie Cloutier[3] and Brianna N. Gaskill[1]*
*[1]Purdue University, Department of Animal Sciences, 270 S. Russell Street, West Lafayette, IN 47907, USA, [2]Purdue University, Center for the Human-Animal Bond, Department of Comparative Pathobiology, 725 Harrison Street, West Lafayette, IN 47907, USA, [3]Independent Scientist, 190 O'Connor St, Ottawa, Canada; lafollet@purdue.edu*

Laboratory rats may experience distress during handling which can reduce rat welfare. Rat tickling, a handling technique that mimics aspects of rat rough-and-tumble play, has been found to induce positive affect based on production of 50-kHz ultrasonic vocalizations (USVs). However, current protocols for rat tickling are time-intensive, making its practical implementation difficult. Our objective was to identify a time-efficient and effective dosage of rat tickling. We hypothesized that affect and handling could be improved by small, daily doses of tickling within a 5-day work week. Long-Evans rats (n=72) of both sexes, housed in pairs were sampled. Each cage was randomly assigned a tickling duration (15, 30 or 60 s per rat) and frequency (1, 3, or 5 days). After the final day of tickling, rats were tested for ease of, and reaction to, handling via an intraperitoneal injection of saline following a tickling session for their assigned duration. On test day, we measured USVs, home cage behavior (60 min before/after testing), approach behavior (30 s before/after testing), and fecal corticosterone. Periods before and after testing measured anticipatory and reactionary responses, respectively. Behaviors included play, activity, location, and indicators of fear or anxiety such as rearing and contact with the hand. Data were analyzed using general linear models. Regardless of duration, rats tickled for 3 days produced a higher rate of 50-kHz USVs during tickling ($F_{1,17}$=28.1, $P<0.001$), played more and were less inactive in their cage for the hour *before* tickling and injection ($F_{1,22}$=10.4, $P=0.004$; $F_{1,44}$=13.5, $P<0.001$), and spent less time in their huts in the cage for the hour *after* injection ($F_{1,126}$=16.7, $P<0.001$) than rats only tickled for 1 day. Additionally, rats tickled for 3 or 5 days reared more and spent more time close to the hand in the approach test after injection ($F_{2,66}$=5.4, $P<0.001$; $F_{2,53}$=5.6, $P=0.02$) than rats tickled for only 1 day, regardless of duration. There were no differences in any outcomes between tickling for 3 or 5 days. Fecal corticosterone was unaffected by either tickling duration or frequency. In summary, we found that tickling duration did not alter any measures and that a 3-day tickling frequency was more efficient and effective than 1-day but equivalent to 5-day, based on increased 50-kHz USVs (a measure of positive affect), positive anticipatory behavior, including play, and positive reactionary behavior. Therefore, we conclude that a time-efficient and effective rat tickling dosage is 15 s for 3 days before any potentially aversive procedures are applied. Overall, our results suggest that minimal rat tickling can be effective in habituating rats to handling and, thus, preparing them for research procedures.

## Odour conditioning of positive affective states: can rats learn to associate an odour with tickling?

Vincent Bombail[1], Nathalie Jerôme[1], Ho Lam[1,2], Sacha Muszlak[1], Simone L. Meddle[2], Alistair B. Lawrence[2] and Birte L. Nielsen[1,3]

[1]Neurobiology of Olfaction (NBO), INRA, Université Paris-Saclay, France, [2]The Roslin Institute, University of Edinburgh, The Royal (Dick) School of Veterinary Studies, Scotland's Rural College (SRUC), United Kingdom, [3]Systemic Modeling Applied to Ruminants (MoSAR), INRA, Université Paris-Saclay, France; birte.nielsen@inra.fr

Rodents can learn to associate a neutral stimulus with a negative experience, such as an electric shock. Few attempts have been made to create associations with positive experiences. In rats, tickling (i.e. mimicking rough-and-tumble play with an experimenter's hand) has been shown to elicit 50 kHz ultra-sonic vocalisations (USVs), which have been compared to laughter in humans. We investigated if young rats could learn to associate the presence of an odour with the putatively positive experience of being tickled. Male Wistar rats (n=24) were pair-housed, and habituated to an inverse lighting schedule as well as being transported to a transparent test arena with sawdust litter and an empty container. All rats were placed in the arena on each testing day (10 days in total), and one of the odours (A and B) were present on alternate days. The rats were subjected to one of 3 treatments (n=8) under red light: never tickled (CON), or tickled, only when odour A (TA) or odour B (TB) was present in the container. After 1 min in the arena, tickling began and consisted of four 20 s episodes of tickling interspersed with 3 pauses each of 20 s where the hand was motionless on the side of the arena, which was also where the hand was for CON and the non-tickled rats of the day. USVs were quantified, and tickled rats produced significantly ($P<0.05$) more USVs when being tickled (mean 233±9 USVs/min) compared to the days when not being tickled (mean 59±7), the latter not differing from the USVs produced by the control rats (mean 29±8). The level of anticipatory USVs in the minute prior to tickling did not differ significantly among the TA, TB and CON rats (range: 40-69). After the 10-day conditioning, we carried out a 3-odour test in the same arena and set-up, but with no tickling or hand present. Rats were placed in the arena for 30 s, and then exposed to 3 odours for one minute each; with 30 s pauses in between. The 3 odours were an unknown odour, extract of fox faeces, and either odour A (for the TA and half of the CON rats) or odour B (for the TB and remaining half of the CON rats). USVs decreased gradually across the testing period, but increased above that of controls for the tickled rats when exposed to their tickling odour (TA: 41±8, TB: 66±11, CON: 20±7 USVs/min), although this only reached significance for odour B ($P<0.01$). Although the rats did not produce more anticipatory USVs when exposed to their tickling odour, they did increase the USV frequency when the odour was presented following two unknown odours, one of which was assumed to be fear inducing. This finding indicates that the rats had learned to associate an odour with the positive experience of tickling. This was not simply because the odour was known, as the control rats showed no such increase. Positive odour conditioning may thus have potential to be developed further with a view to replacing negative odour conditioning tests.

## Chickens play in the wake of humans

Ruth C. Newberry[1], Judit Vas[1], Neila Ben Sassi[2], Conor Goold[3], Guro Vasdal[4], Xavier Averós[2] and Inma Estevez[2,5]
[1]Norwegian University of Life Sciences, Faculty of Biosciences, Dept. of Animal & Aquacultural Sciences, Box 5003, 1432 Ås, Norway, [2]Neiker-Tecnalia, Dept. of Animal Production, 01080 Vitoria-Gasteiz, Spain, [3]University of Leeds, School of Biology, Leeds LS2 9JT, United Kingdom, [4]Norwegian Meat & Poultry Research Centre, Box 396, 0513 Oslo, Norway, [5]Ikerbasque, Basque Foundation for Science, Maria Diaz de Haro 3, Bilbao, Spain; ruth.newberry@nmbu.no

There is little mention of play in literature on the domestic fowl. We explored housing and management factors associated with play behaviour in broiler chickens, predicting that play would occur at higher rates in flocks housed at lower densities, and be stimulated by open space temporarily created in the wake of people walking through the flock. Further, we investigated relationships between play behaviour and measures of health and productivity. We used a novel "behaviour transect" method to sample behaviour in 30 Norwegian commercial broiler flocks at ~28 days of age. We performed 15-s scans to collect data on number of birds engaged in play behaviour (worm running, play fighting, jumping, wing flapping and running, performed spontaneously in an exuberant yet relaxed manner) in ~96 observation patches per flock. The patches were distributed along transects bordered by adjacent feeder and drinker lines, with length demarcated by sets of three consecutive feeder pans. Two observers each sampled patches in the transect located two transects to the left of their current position, one observing undisturbed patches and the other observing disturbed patches vacated 75 s earlier by the first observer. Observers alternated between observing undisturbed and disturbed patches in successive transect walks. Data (number of birds playing per patch offset by estimated total number present) were analysed in a hierarchical Bayesian regression model, with a negative binomial distribution, log link and Markov Chain Monte Carlo estimation. Here, we present mean predicted changes in play behaviour due to significant parameter effects, inferring significance when the 90% highest density interval (HDI) did not cross zero. Play occurred 5.53 (HDI: 5.16, 5.87) times more frequently in disturbed than undisturbed patches (mean predicted number playing/100 birds/15 s: 0.32 vs 0.06), with higher flock densities having a less negative effect on play in disturbed than undisturbed patches (density × disturbance, 10% change in slope; HDI 4, 17). More play was observed in flocks with higher mean body weight (19% change; HDI 6, 31), and lower mortality (-20%; HDI -30, -10), whereas no significant associations between play and feed conversion ratio, incidence of lameness, photoperiod schedule, windows, underfloor heating or feeding programme were detected. Nor did we detect significant effects of methodological parameters such as distance of observation patches from the front or side walls of the house. Wing flapping, running and play fighting made the greatest contributions to the overall rate of play. The disturbance-induced elevation in play levels, especially in higher-density flocks, is indicative of a rebound response to temporarily increased space availability. The association of play with measures of flock vigour supports play as an indicator of positive welfare in broilers.

## Does the motivation for social play behaviour in dairy calves build up over time?

*Maja Bertelsen and Margit Bak Jensen*
*Aarhus University, Department of Animal Science, Blichers Allé 20, 8830 Tjele, Denmark; maja@anis.au.dk*

Dairy calves are often housed in individual pens, which limits play behaviour. A rebound effect is seen following a period with little opportunity to perform locomotor play. This indicates a build-up of motivation to perform locomotor play and preventing this may be a welfare issue. The current study assesses whether the motivation to perform social play also builds up over time. Sixteen mixed-sex pairs of Holstein Friesian calves (avg. age: 5.7 (±0.4) weeks) were housed in pair pens (8.4 m$^2$) for one week and then separated into adjacent single pens (4.2 m$^2$). Over a two-week period, in a cross-over design, the pairs were given access to a deep-bedded arena (3.6×7.10 m) for 45 min on scheduled days such that on test days the pairs had either been deprived of arena access for four days or zero days (the two levels of the experimental treatment). On test days calves' behaviour in the arena was video recorded. Males were chosen as focal calves and observed by continuous recording; females were referred to as companion calves. Health was scored for all calves on test days. The durations of play behaviours were analysed using a mixed model including the fixed effects treatment, week of experiment, order of treatments and the linear effects of focal and companion calf health score. Random effects included block and calf. When deprived of arena access for four days, calves performed more parallel locomotor play than when deprived for zero days (47.0±12.4 s vs 21.4±12.3 s; $F_{1,11}=10.04$, $P<0.01$). Individual locomotor play was not affected by deprivation, but calves tended to perform more individual locomotor play with increasing (worse) companion calf health score ($F_{1,11}=3.72$, $P=0.08$). Calves performed more frontal pushing (social play) at the age of seven than six weeks (back transformed estimates and confidence limits; 89.17 (55.65;130.42) s vs 40.75 (19.01;70.72) s; $F_{1,11}=17.21$, $P<0.001$). Further, a higher health score of the companion calf was associated with a lowered duration of frontal pushing ($F_{1,11}=5.25$, $P<0.05$). The results show that the motivation to perform locomotor play builds up during four days without arena access and that the increased locomotor play behaviour is performed in a social context. Whether parallel locomotor play may be viewed as social play needs further investigation. Frontal pushing, which is social play, was higher in week seven, but no indication of a build-up over time was found. However, the association between the health of the companion calf and frontal pushing may suggest that a healthy calf stimulates more social play. In conclusion, a rebound of parallel locomotor play, but not frontal pushing, suggests that the motivation for some, but not all, elements of calf play behaviour builds-up over time. Play has been suggested to indicate positive welfare, but differentiating between elements of play may be important in the use of calf play behaviour to assess animal welfare.

## The Canadian commercial seal hunt

*Pierre-Yves Daoust*
*Atlantic Veterinary College, University of Prince Edward Island, Pathology & Microbiology, 550*
*University Avenue, Charlottetown, C1A 4P3 Prince Edward Island, Canada; daoust@upei.ca*

Harp seals (*Pagophilus groenlandicus*) have been hunted commercially in eastern Canadian waters for their fur and oil for more than 250 years. Much has been said and written in past decades about animal welfare issues related to this hunt. In the late 1970s and early 1980s, members of the Canadian Veterinary Medical Association became involved in assessing these issues. However, the vast majority of their observations were in the form of internal reports that were difficult to access and lacked specific data. The hunt for whitecoats (recently born harp seals) has been prohibited in Canada since 1986. Young harp seals approximately 1-3 months old, that have been weaned and left by their mother at approximately 10 days of age and have completely shed their white fur, are now the main target of the hunt. Since the early 2000s, a more robust effort has been made by Canadian veterinarians to observe, document and report all aspects of the hunt, with the objective of identifying where and how different components of the hunt could potentially be improved. In six different hunting seasons between 1999 and 2009, a total of 885 young harp seals (live animals or carcasses) harvested during the hunt were observed by four different veterinarians. Potential or obvious issues of poor animal welfare were observed in 1.8 to 5.4% of these seals. Some international working groups were also created to review the Canadian seal hunt. Reports from these groups concluded that the tools (club, hakapik, rifle) and methods used in that hunt were adequate to kill the majority of the animals quickly, but that improvement might be needed in the implementation of their use in order to minimize negative welfare outcomes. Part of this work led to the design of a simple 3-step process (stunning, skull palpation, bleeding) to be used consistently by sealers and aimed at ensuring that the vast majority of seals were killed as quickly as possible. The need to perform this 3-step process was included in the "Conditions of License" for the Canadian harp seal hunt in 2009. So far, most of this work has focussed on the hunt for young harp seals, and based on the information available, this hunt compares well with other types of hunt. However, new initiatives have since been taken by the sealing industry to diversify its market, such as extraction of omega-3 fatty acids from the blubber and commercialisation of meat products. Thus, different age groups (adults) and different species (grey seals, *Halichoerus grypus*) are now being targeted in addition to the traditional hunt for young harp seals. This has raised new potential challenges pertaining to animal welfare, such as the use of the hakapik for the bigger young grey seals and the use of rifles of different calibres for adult seals. These are currently being addressed with the sealing industry's collaboration, an industry that continues to be of great economic importance to coastal communities in the region.

**Animal welfare complementing or conflicting with other sustainability issues**

*Donald M. Broom*
*University of Cambridge, Veterinary Medicine, Madingley Road, Cambridge CB3 0ES, United Kingdom; dmb16@cam.ac.uk*

Systems for the production of food, or other product for human use, should be sustainable. This means that the system should be acceptable now and its expected future effects should be acceptable, in particular in relation to resource availability, consequences of functioning and morality of action. However, there are many components of sustainability. People who consider only one aspect may not advocate the best solution. If the focus is entirely on animal welfare, some other harm may be done. Similarly, focus entirely on preservation of rare wildlife species, maximising local biodiversity or minimising greenhouse gas production may harm welfare. When an agricultural or other product is considered, life cycle analysis of the product takes account of every contributory factor. Every externality of the system should be evaluated and the value of each balanced. Some actions that improve animal welfare may also have positive environmental effects and each aspect can be measured. If straw from cereal production is burned, carbon dioxide is released into the atmosphere but if straw is used as bedding or for manipulation to benefit farm animals and then composted, the greenhouse gas effect is much less. If wild animals are kept as pets, for most species their welfare is very poor and for some species wild populations are reduced. Stray dogs may have a large negative impact on the populations and welfare of some wild animals and their welfare is often poor because of disease and malnutrition so humanely killing the dogs can prevent poor welfare and benefit conservation. If cattle grazed on extensive fertilized pasture are moved to semi-intensive silvopastoral systems, the presence of shrubs and trees greatly increases biodiversity, reduces greenhouse gas production per unit of production, reduces conserved water usage and improves welfare because of: shelter from the sun, increased habitat choice and reduced disease resulting from more predators of tick and fly disease vectors. On the other hand, conserving land for hunting wild animals increases biodiversity but the hunting usually causes poor welfare. The land-sparing argument, encouraging intensive animal production so more land is available for nature reserves, would favour feedlots for beef production but the welfare of the cattle in feedlots is often poor. Where endangered species cannot adapt well to captive conditions, captive breeding might preserve the species but the welfare of the animals might be poor. The approach advocated here is that, when a system is being evaluated, each of the many components of sustainability should be measured precisely: welfare, biodiversity, worker satisfaction, water use, greenhouse gas production and harmful accumulation of pollutants like nitrogen and phosphorus. Decision-making may involve comparing the extent of each positive and negative consequence or considering any negative that is so great that no counter-balancing would ever be acceptable to the public.

## Impacts of population management with porcine zona pellucida vaccine on the behaviour of a feral horse population

*Katrina Merkies[1,2], Abbie Branchflower[1,2] and Tracey Chenier[3]*
[1]*University of Guelph, Campbell Centre for the Study of Animal Welfare, 50 Stone Rd E, N1G 2W1 Guelph, Canada, [2]University of Guelph, Animal Biosciences, 50 Stone Rd E, N1G 2W1 Guelph, Canada, [3]University of Guelph, Population Medicine, Ontario Veterinary College, 50 Stone Rd E, N1G 2W1 Guelph, Canada; abranchf@uoguelph.ca*

Removing feral horses from rangelands has been the primary form of population control, however removals pose welfare concerns. On-range management using porcine zona pellucida vaccine (PZP), a female contraceptive, is an alternative, but PZP use has been indicated in the disruption of social dynamics within bands. The objective of this study was to quantify sexual and aggressive behaviours by and toward PZP-vaccinated mares. Fourteen focal bands (n=83 horses) meeting the criteria of a stallion and at least two sexually mature mares with or without offspring were observed in the Pryor Mountain Mustang Reserve, Montana, USA. Age, parity, pregnancy and PZP status were obtained from the records of the local Bureau of Land Management office. Observations were conducted in 30-min windows in the mornings and afternoons of 17 days within a five-week period during June and July 2017. The mare closest in proximity to the band stallion was recorded at each minute, and all occurrences of sexual (penile erection, genital sniffing, mounting, flehmen, marking) and aggressive (bite, kick, threats, snaking, chasing) behaviours were recorded. Mares were categorized into two treatment groups: PZP+ (vaccinated with PZP in 2016), or PZP- (not vaccinated with PZP in 2016) and Kruskal-Wallis test by rank was used to determine differences in the sexual and aggressive behaviours received or initiated by the mares based on PZP status. A general linear mixed model determined the mare found most often in close proximity to the band stallion. Pearson correlation determined any relationship between the stallion behaviours of flehmen and marking. Evaluation of proximity data demonstrated that PZP+ mares were more often found closer to the stallion than PZP- mares (M=29.7%, 95% CI[26.3, 33.1] vs M=23.1%, 95% CI[19.5, 28.4] respectively, $P<0.042$), however this was not constrained to any particular mare ($P>0.055$). There were no differences in any sexual or aggressive behaviours received ($P>0.21$) or initiated ($P>0.07$) by PZP+ vs PZP- mares. There was no correlation between the stallion behaviours of flehmen and marking ($P>0.98$). Treatment of mares with PZP vaccine to prevent pregnancy did not appear to alter the sexual or aggressive behaviours either directed to them or initiated by them toward the band stallion. Although treated mares continue to cycle throughout the reproductive season this does not appear to influence the social dynamics within the band. Mares treated with PZP are more often found in closer proximity to the band stallion due to the stallion being more vigilant toward PZP+ mares, increased interest on the part of the mare, or the mare seeking protection from external harassment from other stallions. As no disruptive effects of PZP treatment on the social ecology of the Pryor Mountain mustangs were evident in this study, the use of PZP vaccine for population control remains a viable solution and likely maintains better welfare than removals.

### Why does annual home range predict stereotypic route-tracing in captive Carnivora?

*Miranda Bandeli[1], Emma Mellor[2] and Georgia Mason[1]*
*[1]University of Guelph, Animal Biosciences, 50 Stone Road East, Guelph, ON, N1G 2W1, Canada,*
*[2]Bristol University, School of Veterinary Sciences, Langford House, Langford, Bristol, BS40 5DU,*
*United Kingdom; miranda.bandeli@gmail.com*

In captivity, some Carnivora species respond well, while others are prone to high levels of stereotypic behaviour (mainly route-tracing, RT, e.g. pacing). Previous work by found that species with naturally large annual home ranges (AHR) show higher levels of RT (median % time spent RT by stereotypic subjects). This effect was robust, although never very strong. To understand its underlying drivers, and perhaps find stronger predictors, we asked whether any potential determinants of natural AHR (e.g. metabolic rates, population density, & predation risk), or any potential consequences of wide-ranging lifestyles (e.g. long daily travel distances & small ratios of daily distance to annual distance travelled [DD:AD]) are better predictors of RT than AHR itself. We updated Kroshko *et al.*'s RT database, to contain data from 2,337 subjects across 56 species (28 being well-enough sampled to include in analyses, using a minimum inclusion value of 5 subjects per species: cf. Kroshko *et al.*). Literature searches yielded median natural AHR values for 23 of these species. Phylogenetic generalized least square regressions confirmed that wide-ranging species spend more time in RT ($P=0.012$, $F_{1,21}=5.98$, $R^2=0.185$). Via literature search, we then collected data on the correlates of natural AHR. Added into our statistical models, none of these variables eliminated the AHR effect on RT, so none explained why AHR predicts RT. Daily distance travelled & habitat productivity did not predict RT; while body mass, individual metabolic need & population density did not predict RT when controlling for AHR. Unexpectedly, however, combining two variables with AHR greatly improved the model $R^2$: non-predated, wide-ranging species perform higher levels of RT (whole model: $P=0.0022$, $F_{2,10}=11.93$, $R^2=0.65$; AHR term only: $P=0.013$; predation levels only: $P=0.004$), as do wide-ranging species with small DD:AD ratios (whole model $P=0.0047$, $F_{2,16}=7.62$, $R^2=0.42$; AHR term only: $P=0.029$; DD:AD term only: $P=0.022$). Thus since AHR was not eliminated by any of our variables, it remains unclear why AHR predicts RT across captive Carnivora; more research is needed here. However, we did newly find that non-predated, wide-ranging Carnivora with small DD:AD ratios are most at risk for high levels of RT. It may be that prey species spend more time engaging in predator-avoidance, e.g. hiding, with this reducing their RT. The effects of DD:AD ratio, in contrast, may indicate the importance for RT-vulnerability of being from naturally semi-nomadic species, with more variety in their daily lives (compared to species that naturally cover much of their annual range each day). We hope this study helps zoos to design more appropriate enclosures and enrichments (e.g. offering daily choices of different enclosures to explore), and to predict which Carnivora species are likely to be most at risk of high RT.

**Effects of manipulating visitor proximity and intensity of visitor behaviours on little penguins (*Eudyptula minor*)**

*Samantha J. Chiew[1], Kym L. Butler[1], Sally L. Sherwen[2], Grahame J. Coleman[1], Kerry V. Fanson[3] and Paul H. Hemsworth[1]*
[1]*Animal Welfare Science Centre, The University of Melbourne, 21 Bedford St, Level 2, North Melbourne VIC 3051, Australia, [2]Department of Wildlife Conservation and Science, Zoos Victoria, Elliott Ave, Parkville VIC 3052, Australia, [3]Centre for Integrative Ecology, Deakin University, 75 Pigdons Road, Waurn Ponds VIC 3216, Australia; schiew@student.unimelb.edu.au*

Limited research has been conducted to understand the effects of visitors on zoo penguins despite the growing evidence showing negative effects of human disturbance on wild penguin populations. This experiment examined the effects of manipulating the proximity of visitors and the intensity of visitor behaviours on fear behaviour of 15 little penguins (*Eudyptula minor*). A 2×2 factorial fully randomised design, with an added control where the exhibit was closed to visitors, was used to examine 2 main factors at 2 levels: (1) viewing proximity of visitors to enclosure: 'Normal viewing proximity' and 'Increased distance from enclosure' (using a barrier 2 m from the enclosure); and (2) intensity of visitor behaviours: 'Unregulated' and 'Regulated' (using signage). Treatments were imposed for 2-day periods with 3 replicates of each treatment (total of 30 study days). Behavioural observations using CCTV footage were conducted, using a combination of instantaneous point sampling and one-zero sampling to record penguin behaviour including huddling, proximity to the visitor viewing area, vigilance, surface swimming and preening. Direct observations were conducted using instantaneous point sampling to record visitor number and continuous sampling to record visitor behaviours that were identified to likely disrupt penguins including banging on enclosure structures, shouts/screams, leaning over the enclosure, tactile contact with the pool water and sudden movement. Penguin faecal glucocorticoid metabolites (FGM) were analysed as a measure of stress physiology. Data were analysed using the general analysis of variance for a fully randomised factorial design, with an added control. Increasing visitor distance from the enclosure reduced all visitor behaviours except shouts/screams. When visitors were further away, there was a reduction in the proportion of penguins huddling ($P=0.0057$), idle ($P=0.040$) and vigilant ($P=0.048$) and an increase in the proportion of penguins within 1 m of the visitor viewing area ($P=0.0039$), surface swimming ($P=0.0028$) and preening in the water ($P=0.020$). Visitor intensity effects were relatively limited. No interaction was found between the treatments ($P>0.05$) and no effect of treatment on penguin stress physiology ($P>0.05$) was found. These results indicate that visitor proximity affects penguin avoidance responses, but there was no prolonged change in stress physiology that would suggest a chronic stress response. Thus, close visitor contact can be fear-provoking for little penguins, but a physical barrier to increase the distance between visitors and penguins can reduce their avoidance of visitors. Further research is required to determine whether it is proximity per se and/or specific visitor interactions such as leaning over the enclosure or tactile contact with the pool that elicits penguins' avoidance responses.

## Verification of the effectiveness of environmental enrichment and the influence of visitors on captive tigers

*Momoko Oka[1], Yumi Yamanashi[2], Kota Okabe[2], Masayuki Matsunaga[2] and Satoshi Hirata[1]*
*[1]Wildlife Research Center of Kyoto University, 2-24 Tanaka-Sekiden-cho, Sakyo, Kyoto, 606-8203, Japan, [2]Kyoto city Zoo, Okazakihousyouzimati, Sakyo, Kyoto, 606-8333, Japan; momo950322@yahoo.co.jp*

Zoo animals are often bred in environments that are significantly different from their original habitats. As a result, abnormal behaviors such as pacing around in the exhibit, shaking the body in front and back, left and right, coprophagy and regurgitation may occur. Among them, pacing is common in large felid species such as tigers. Environmental enrichment can be used for improving the welfare of zoo animals and is considered important for decreasing pacing. Several factors such as the presence or absence of enrichment and the influence of visitors are involved in pacing, but such factors are rarely investigated simultaneously. In this study, we investigated the effectiveness of enrichment and the factors influencing the behavior of captive Amur tigers (*Panthera tigris altaica*). We observed the behavior of 3 amur tigers in the Kyoto city Zoo and examined the effect of types of enrichment, ambient temperature and number of visitors on behavior. The results indicated that there are individual differences in the preference for enrichment. Moreover, as the maximum temperature gets higher, the pacing frequency decreases and the resting frequency tends to increase. The influence of visitors was not as great as that of the ambient temperature. Ambient temperature is an important factor influencing the behavior of tigers. Since there are individual differences in the preference for enrichment, providing multiple types of enrichment is useful for tigers.

## Can surface eye temperature be used to indicate a stress response in seals (*Phoca vitulina*)?

*Amelia Macrae and David Fraser*
*University of British Columbia, Animal Welfare Program, 2357 Main Mall, Vancouver, V6T 1Z4, Canada; amarimacrae@gmail.com*

Many mammalian species (e.g. mice, cattle, horses) demonstrate a change in eye temperature in response to stressful, and possibly to painful routine procedures. Although the physiological basis for the changes is not fully understood, it has been suggested that such changes reflect autonomic nervous system and hypothalamic-pituitary axis (HPA) activity and thus have been proposed as a useful index of a stress response. Infrared thermography (IRT) is increasingly being used to measure physiological stress responses in animals via changes in eye temperature, as this method can indicate a stress response without invasive sampling or handling. The aim of this study was to determine whether the eye temperature of harbour seals (*Phoca vitulina*) changes in response to routine handling (capture and restraint). Fifty-two pups (healthy, ~90 d old) admitted to a Canadian rehabilitation facility had their eye temperature recorded (measured with FLIR T300 IRT camera, images recorded approximately every 10 s) during either a period of no handling followed by a single handling session or two repeated handling sessions. All pups first had baseline eye temperatures recorded for 3 min before any handling began. Then seals underwent their assigned treatment: Group 1 (n=26, handled once) had eye temperatures recorded for 3 min during a period of no handling, and then 10 min later were captured and had eye temperatures recorded for 3 min of restraint. Group 2 (n=26, handled twice) were captured and had eye temperatures recorded for 3 min of restraint (prior to receiving an injection), were released for 10 min and then were restrained again for 3 min when eye temperature was recorded. Maximum eye temperature was calculated for each image using FLIR Tools+ software, and images from each 3-min recording period of each pup were then pooled. Data were analyzed using linear mixed effects models in R. Compared to the baseline, eye temperature increased 0.5±0.18 °C (mean ± SE, $P<0.01$) more when pups were handled the first time compared to pups that were not handled. All pups had a significant increase in eye temperature of 0.7±0.09 °C (mean ± SE, $P<0.001$) from before to after the first time being handled. Eye temperature of pups that underwent a second handling event increased a further 0.7±0.08 °C (mean ± SE, $P<0.001$) from the first time they were handled to the second time. The higher eye temperature of handled vs non-handled pups, and the increase from before to after handling, suggest that handling and restraint cause a physiological stress response detectable via IRT. Additionally, the increased temperature seen the second time pups were handled suggests the first handling likely was aversive, resulting in an anticipatory response to their second handling. These results show promise for the use of eye temperature to indicate a stress response and for evaluating routine procedures in seals.

**Piggy in the middle: the cost-benefit trade-off of a central network position in regrouping aggression**

Simone Foister[1], Andrea Doeschl-Wilson[2], Rainer Roehe[1], Laura Boyle[3] and Simon Turner[1]
[1]SRUC, Roslin Institute Building, Easter Bush, EH25 9RG, United Kingdom, [2]University of Edinburgh, Roslin Institute Building, Easter Bush, EH25 9RG, United Kingdom, [3]Teagasc, Animal & Grassland Research and Innovation Centre, Moorepark, Fermoy Co. Cork, P61 C997, Ireland; simone.foister@sruc.ac.uk

During the first 24 hours (24 h) post-regrouping pigs exhibit high levels of aggression in order to establish a dominance hierarchy. Large variation exists in the number of injuries (skin lesions) caused by short and long-term aggression, indicating that certain groups struggle to establish and maintain a stable hierarchy. To date, dyadic measures have not fully explained this variation, suggesting a more refined tool for understanding the behaviour may be required. Here we test the ability of social network analysis (SNA) to predict skin lesions (SL) at 24 h post-regrouping (SL24h) and in stable groups 3 weeks later (SL3wk). Pigs (n=1,170) were mixed at 8 wks into single-sex pens of 15 and video recorded for 24 h. All aggressive interactions were included to create a network for each pen. A stepwise regression identified the size of cliques (fully-connected sub-groups) and betweenness centralisation (poor connectivity) as the network structures that provided the best model fit for pen level SL residuals. Full model fit statistics were obtained using REML mixed models with experimental batch as a random effect; sex, breed, mean body weight, and network structure (clique size/betweenness) as fixed effects, and pen level SL as the response variate. To assess how individual position within such structures related to individual SL, animals were categorised as either 'clique-members' or 'non-members', and 'high' or 'low' betweenness centrality (individuals that connect otherwise unconnected animals). The effect of individual network positions was analysed by entering batch and pen as a random effect; sex, breed, body weight, and network position as fixed effects, with individual SL as the response variate. At the pen level, high betweenness centralisation at regrouping led to higher SL3wk than low betweenness centralisation ($F_{1,57}$=4.38, $P<0.05$). At the individual level, pigs with high betweenness centrality had lower SL24h than low centrality pigs ($F_{1,978}$=3.95, $P<0.05$). Yet there was no significant difference in SL3wk ($F_{1,976}$=0.06, $P$=0.81), revealing all animals had equally high SL regardless of their centrality. In contrast, pens that formed large cliques at regrouping had lower SL3wk ($F_{1,58}$=6.25, $P<0.01$) than pens with smaller cliques. Clique-members had higher SL24h than non-members ($F_{1,978}$=9.19, $P<0.01$), and there was no significant difference in SL3wk ($F_{1,978}$=0.28, $P$=0.59) demonstrating that all animals in the pen had similarly low SL regardless of their clique membership. We conclude that early network structures offer predictive value for long-term aggression. Betweenness centralisation appears to increase long-term aggression, whereas large cliques appear to aid hierarchy formation and lead to low injury rates for all pen members at 3 weeks while not causing excessive injury at regrouping.

## Preferential associations in recently mixed group-housed finisher pigs

Carly I. O'Malley[1], Juan P. Steibel[1,2], Ronald O. Bates[1], Catherine W. Ernst[1] and Janice M. Siegford[1]
[1]Michigan State University, Animal Science, 474 S. Shaw Lane, East Lansing, MI 48824, USA, [2]Michigan State University, Fisheries and Wildlife, 480 Wilson Road, East Lansing, MI 48824, USA; omalle50@msu.edu

A common practice in modern pig farming is to mix unfamiliar pigs together at different production stages in order to create uniform groups based on weight and sex. Consequently, pigs show high levels of aggression for 24-48 h after mixing as they work to establish a social hierarchy. To reduce levels of aggression at mixing, we need to better understand how pigs integrate into new social groups. In this study, we explored social integration in five pens of finisher barrows (castrated males) to investigate if pigs show preferential associations with familiar over unfamiliar pigs. Pigs were mixed into new groups at 10 wk of age, based on weight, into finisher pens (4.83×2.44 m). Each pen contained 10-15 pigs (67 pigs total), made of groups of 2-5 familiar pigs. Familiar pigs were defined as pigs housed together in nursery pens immediately prior to mixing into finisher pens. A small number of pigs were siblings that had been housed together only when the litter was with the dam (n=21). The remaining pigs in each pen were completely unfamiliar with one another. Pigs were video recorded for 4 consecutive days immediately after mixing. Videos were decoded using 5 min scan-sampling between 14:35-16:35 for each day (4 samplings total). Association matrices were created using the half-weight association index (HWI) for each dyad in physical contact without overt aggression. A HWI score ≥0.15 (double the mean) indicated a preferential association. Independent linear models were fitted to HWI data within each observation day. Models included fixed effects of pen (4 levels), previous penmate status (1 if a dyad was composed of nursery penmates, 0 if a dyad was composed of unfamiliar pigs) and littermate status (1 if the dyad was composed of siblings and 0 if the animals in the dyad were not siblings). Significance testing was performed using an exact permutation test. Among nursery penmates, 47% had HWI≥0.15 (preferential association) and 53% did not. Among unfamiliar pigs, only 9% had HWI≥0.15 and 91% did not (P<0.01). On day 2, 25% of associations between nursery penmates and 17% for unfamiliar pigs were preferential (P=0.043). On day 3, 20% of associations between nursery penmates and 20% for unfamiliar pigs were preferential (P>0.05). On day 4, 20% of associations between nursery penmates and 18% for unfamiliar pigs were preferential (P>0.05). The pigs did not show preferential associations with siblings that had been housed together before weaning (P>0.05), and there was no effect of pen (P>0.05). These results suggest that immediately after mixing, pigs remained in close physical contact with familiar conspecifics but integrated into the new social group within 50 h of mixing. Group-housed pigs show a reduction in aggression within 48 h of mixing, which may coincide with social integration. Investigating factors that promote social integration could, therefore, be useful in reducing aggression at mixing.

## Estimation of genetic parameters of agonistic behaviors and their relation to skin lesions in group-housed pigs

*Kaitlin Wurtz[1], Juan Steibel[1,2], Carly O'Malley[1], Ronald Bates[1], Catherine Ernst[1], Nancy Raney[1] and Janice Siegford[1]*
*[1]Michigan State University, Animal Science, 474 S Shaw Ln Room 1290, East Lansing, MI 48824, USA, [2]Michigan State University, Fisheries and Wildlife, 480 Wilson Rd #13, East Lansing, MI 48824, USA; wurtzkai@msu.edu*

Aggression in group-housed pigs following mixing with unfamiliar animals is a pressing welfare and production issue in the swine industry. Genetic selection for pigs better suited for group living has been proposed. However, a better understanding of the underlying genetic components of aggression and their relationship to easily measured phenotypic traits is necessary before widespread adoption of such a strategy into breeding programs. Our study used skin lesion counts (fresh, bright red cuts) as a readily measurable proxy for assessing level and type of aggression. We estimated genetic and phenotypic correlations of lesion scores with duration of aggressive behaviors obtained through video observation. Observations were obtained on 393 purebred Yorkshire pigs at the grow-finish stage (~67 d of age) immediately following mixing into new pens containing 13-15 pigs (0.79 to 0.98 m$^2$/pig) with unfamiliar animals (3 to 5 familiar pigs) and on 498 individuals 3 wk following remixing to observe aggression after stable hierarchies would supposed to be formed. To examine phenotypic and genetic correlations between behaviors and lesions, bivariate analyses were performed using genomic best linear unbiased prediction models. Fixed effects were composed of sex and replicate, covariates were composed of weight and pre-mixing lesions, and pen and genetic additive effect were evaluated as random effects. The response variables were log-transformed post-mixing lesions and log-transformed duration of time spent engaging in specific aggressive behaviors. A genetic relationship matrix was constructed using genotypes from 50,924 single nucleotide polymorphisms. Univariate models with the same model components were fitted for each behavior to estimate heritabilities. Both inverse and parallel pressing were highly correlated with reciprocal fights, and thus were grouped for further analyses. Heritability estimates of behaviors were generally low (avg. 0.09) at mixing and at the 3 wk post-mixing period. However, heritability estimates were larger (avg. 0.13) when behaviors were categorically grouped (i.e. reciprocal interactions, delivering aggression, receiving aggression). At mixing, reciprocal aggressive interactions were genetically ($r_g$) and phenotypically ($r_p$) positively correlated with lesions to the anterior ($r_g$=0.74, $r_p$=0.39), central ($r_g$=0.43, $r_p$=0.30), and caudal ($r_g$=0.42, $r_p$=0.25) regions of the body. Delivery of one-sided aggression was positively correlated with anterior lesions ($r_g$=0.55, $r_p$=0.16). Genetic and phenotypic correlations 3 wk post-mixing were close to zero and not significant, potentially due to small population size and low numbers of lesions observed at this time. Based upon these preliminary findings, it appears anterior skin lesion counts show the greatest potential for reducing aggressive behaviors though selection. This knowledge will help guide genetic selection by determining the traits to select for optimal change.

## Pen-level risk factors associated with tail lesions in Danish weaner pigs: a cross-sectional study

*Franziska Hakansson, Anne Marie Michelsen, Vibe P. Lund, Marlene K. Kirchner, Nina D. Otten, Matthew Denwood and Björn Forkman*
*University of Copenhagen, Department of Veterinary and Animal Sciences, Groennegaardsvej 8, 2400, Denmark; fh@sund.ku.dk*

Tail-biting in pigs inflicts severe pain, reduces the welfare of the bitten pig and can lead to infections, potentially resulting in a systemic disease and therefore economic losses for the farmer. Reasons for performing tail-biting behaviour are multifactorial and motivations are thought to range from: increased need for explorative foraging and oral manipulation, limited availability of resources leading to aggressive competition and reduced coping ability attributable to poor health or augmented stressful events happening in crucial periods. As part of a cross-sectional project, this study investigated the association between potential risk factors and the occurrence of one or more pigs per pen with tail lesions on 19 commercial pig herds. Herds were sampled based on a minimum of 500 growing pigs (mean: 2,219 pigs) and voluntary participation, and were visited once during 2015. Random sampling of pens and animals and data collection was based on the latest Welfare Quality® protocol for pigs. A total of 2,328 pigs up to 50 kg (average weight: 22 kg; average no. pigs in pen: 30) from 158 pens were assessed and pen parameters collected. All pigs were tail-docked. Pen-level risk factors included in the study were: number of pigs with ear lesions, manure and wounds on the body, and rooting material, type of feeding system, water-points per pig and stocking density. Associations between any pen-level occurrence of tail-biting (at least one assessed pig per pen with tail lesion) as a binary outcome and selected risk factors were analysed using a multivariable mixed effect logistic regression model with herd as a random effect. In 41 (26%) of the 158 pens assessed, tail lesions were observed. The risk of tail-biting was significantly higher in pens with: low-value rooting material (e.g. chains) compared to high-value material (e.g. straw) (OR=10.7, 95% CI: 3.4-147.4), and in pens where more than 50% of the assessed pigs were dirty (manure on more than 20% of the body) (OR=4.6, 95% CI: 1.8-20.9). There was also an observed increased risk in pens where more than 50% of the assessed pigs had ear lesions (OR=3.9, 95% CI: 1.1-28.4). There were no significant associations for the other parameters assessed. The use of low-value rooting material could lead to redirected oral manipulation and thus, increased risk of tail-biting and ear lesions. Pen hygiene can possibly be affected by several parameters such as overcrowding, lack of substrate and inadequate climatic conditions and management. Hence, taking the dirtiness of a pig as an indicator of poor pen conditions, the results of this study suggest an association between inadequate housing and tail-biting. The lack of an association between wounds and tail lesions in this study contradicts the hypothesis that tail-biting may be related to aggressive competition.

## Tracing the proximate goal of foraging in pigs

*Lorenz Gygax[1,2], Julia Moser[1] and Joan-Bryce Burla[1]*
*[1]FSVO, Centre for Proper Housing of Ruminants and Pigs, Tänikon, 8352, Switzerland,*
*[2]Humboldt-Universität zu Berlin, Animal husbandry, ADTI, Berlin, 10099, Germany;*
*lorenz.gygax@hu-berlin.de*

Pigs often retain a high motivation for foraging even if they are fed ad libitum. Therefore, satiety does not control foraging alone. If the need to forage is not satisfied, behavioural disorders may develop. Consequently, foraging motivation of pigs is of a high concern with regard to animal welfare. We investigated how much the opportunity to perform specific foraging behaviours can satisfy the motivation to forage in dry sows. This idea was based on the assumption that performing some part of foraging behaviour could itself represent a part of the proximate goal that pigs try to achieve. We elicited specific foraging behaviours by providing different substrates: rooting and manipulating stones by "earth materials" (E, soil and stones), biting and chewing by "fibre materials" (F, sisal ropes and birch tree branches), grazing by a "grazer" (G, grass and silage in rack with a narrow wire mesh). All materials included some edible parts. Each of 24 sows went through a three-pen sequence eight times. In each sequence, one substrate was provided in the first two pens. This lead to six trials including every possible sequence of two of the three substrates (balanced across sows). In pen 3, all substrates were provided simultaneously (A). The sows' behaviour in pen 3 was taken to reflect the remaining motivation to perform specific foraging behaviours. As a control, A was provided in all three pens in two additional trials. Sows could stay in each pen up to 40 min, but were moved on if they stopped foraging for 5 min. Mixed-models with parametric bootstrap were used for statistical evaluation. Treatments E, F, G specifically elicited the behaviours as described above (all $P\leq0.001$): e.g. in pen 1, sows rooted the most with E compared to F and G. Moreover, A was more attractive for the sows than a single substrate: they spent a higher proportion of the allocated time in pen 1 when A was provided compared with E, F, or G ($P\leq0.001$). In addition, sows spent the shortest time in the pen 3 after they had A in pen 1 and 2 ($P=0.014$): A (model estimate and [95% confidence interval]): 52% [43-62]; no E: 71% [62-80]; no F: 64% [54-74]; no G: 68% [59-78]. In pen 3, sows increased performing the specific behaviour, which was prevented in pens 1 and 2 (all $P\leq0.03$): e.g. they rooted more if they had encountered F and G before. Although A in pen 1 and 2 reduced foraging motivation in pen 3, no single behaviour elicited in pens 1 or 2 accounted for this and, therefore, acted as a goal state in pig foraging. If the goal state of foraging lay in performing foraging behaviours, a variety of such behaviours or a specific sequence of these behaviours may be necessary to reach the goal. Exploration with its own goal state may also interfere. To conclude, there is no simple behavioural remedy for quenching foraging motivation. For practical application, it still seems that providing different material types might be best suited to satisfy the motivation to perform foraging behaviour.

## Effect of temporary confinement during lactation on the nursing and suckling behaviour of domestic pigs

*Gudrun Illmann, Sébastien Goumon, Iva Leszkowová and Marie Šimečková*
*Institute of Animal Science, Department of Ethology, Přátelství 815, 10400 Prague, Czech Republic; gudrun.illmann@vuzv.cz*

Farrowing pens with temporary crating (TC) have been developed as a compromise between conventional farrowing crates and pens to better accommodate the needs of the sow and of the piglets. The sow is only confined during the first few days post-partum, thus protecting piglets from crushing when they are the most vulnerable. After opening the crate, the increase in the available space may trigger higher activity in sows and may lower their motivation to nurse which may impair piglet weight gain. The increased space in temporary crating pens may also allow a more intensive progressive weaning, as found in other studies in which sows could move away from the piglets, by reducing the number of nursings bouts and increasing nursing terminations by the sows at the end of lactation. The aim of the study was to assess short-term (i.e. 24 h after opening the crate) and long-term effects (day 25) of the removal of confinement on day 3 post-partum on nursing and suckling behaviour compared to permanent farrowing crates. Sows were crated from 5 days pre-partum to either weaning (permanently crated; PC group; n=14) or 3 days post-partum (TC group; n=13). Sows and their litters were observed on days 4 and 25 for 4 hours. Duration of pre- and post-massages (massage of the udder by the piglets before and after milk ejection), nursing termination by the sow, number of piglets missing milk ejection, number of piglets fighting during pre- and post- massages and nursing success (i.e. with or without milk ejection) were noted. Data were analysed using PROC GLM and PROC GENMOD of SAS including housing, litter size and parity as fixed effects. Nursing behaviour did not differ between sows housed in temporary crating pens and those housed in permanent crates on days 4 and 25 post-partum (i.e. same number of nutritive nursings, same proportion of non-nutritive nursings, same duration of post-massages and same proportion of nursing termination of post-massages by the sow). Suckling behaviour was similar between treatments. There was no difference in the number of piglets attending pre-and post-massages, proportion of fighting piglets during the pre-and post-massages and the proportion of piglets missing milk ejection. However, with increasing litter size, sows terminated post-massages more often ($P<0.05$) and piglets had shorter post-massages ($P<0.05$) on day 4. Furthermore, a higher proportion of piglets were missing a milk ejection on days 4 and 25 ($P<0.05$) independent of the housing system. In conclusion, housing had a very limited effect on nursing and suckling behaviour. It indicates that the sow and piglet behaviours are not impaired after opening the crate (short- term effect) and the increased space after opening did not allow a progressive weaning on day 25. Increased litter size significantly impaired nursing and suckling behaviour of the sow and piglets independent of the housing system.

## Effect of design of outdoor farrowing hut on environment inside the hut, and sows and piglets performance and welfare

*Lydiane Aubé[1,2], Frédéric Guay[1], Renée Bergeron[3], Sandra Edwards[4], Jonathan Guy[4] and Nicolas Devillers[2]*
[1]*Université Laval, 2425, rue de l'Agriculture, Québec, Canada,* [2]*Agriculture and Agri-Food Canada, 2000 College Street, Sherbrooke, Canada,* [3]*University of Guelph, 50 Stone Road East, Guelph, Canada,* [4]*Newcastle University, King's road, Newcastle, United Kingdom; nicolas.devillers@agr.gc.ca*

The aim of this study was to determine the impact of outdoor farrowing hut design on performance and welfare of sows and piglets. Fifteen groups of three sows were distributed in 5 blocks. Within each group, sows had access to 3 different farrowing huts: an English-style metal ark (Metal), a round plastic calf hut (Plastic) and a home-made wooden A-frame hut (Wood). Huts differed in their characteristics (e.g. shape, floor area (4.7, 3.7 and 5.2 m², respectively), volume (3.5, 5.7 and 4.4 m³, respectively)). Temperature and humidity in the huts were measured throughout the experiment. Piglets were weighed at d1 of lactation and at weaning (36±5 d). Piglet mortality was monitored until weaning. The environment inside the huts was analysed separately for day and night periods and temperature amplitude (max-min) on a daily basis, according to a generalized complete block design and post-hoc comparisons (Tukey's adjustments of the Student's t-test). Hourly differences between temperatures inside and outside the huts were compared (weighted variance) according to 6 categories of ambient temperatures: <7, 7-12, 12-17, 17-22, 22-27 and >27 °C. Sows did not show a clear preference for a type of hut at farrowing. Daily amplitude of temperature was higher in Plastic (17.9 °C) than in Metal (13.3 °C) and Wood huts (9.9±1.3 °C, $P<0.001$). Throughout the experiment, Plastic huts had a lower average, minimum and maximum temperature during the nights compared to Metal and Wood huts. Humidity during the night was higher in Plastic (89.3%) than in Wood huts (79.2%), Metal being intermediate (84.4±1.1%, $P<0.001$). In contrast, during the light period, humidity in Plastic was lower than in Wood huts (63.3 vs 70.1±1.8%, $P=0.028$). During the light period, when ambient temperature was between 17-22 °C, 22-27 °C and above 27 °C, the difference between ambient temperature and temperature inside Plastic huts was higher than in Metal huts (except for ambient temperature above 27 °C), which was also higher than in Wood huts. For example, above 27 °C, it was 6.2 °C, 3.9 °C and 1.4±0.7 °C warmer in the huts than outside, for Plastic, Metal and Wood, respectively ($P=0.011$). During the night, when ambient temperature was below 7 °C, between 7-12 °C and 12-17 °C, the difference between ambient temperature and temperature inside Plastic huts was lower than in Metal and Wood huts. For example, below 7 °C, it was 3.3 °C, 9.1 °C, and 10.4±0.9 °C warmer in the hut than outside, for Plastic, Metal and Wood, respectively ($P<0.001$). Design of hut did not have any impact on the number of stillborn (5.75±2.1% of total born piglets), mortality rate by crushing (11.3±3.2%), total mortality rate (33.4±5.6%) and piglets' growth rate (ADG: 360±21 g/day). Despite the fact that Wood huts provided a more optimal environment for both piglets (warmer and less humid during cold periods) and sows (cooler during warm periods) than the other hut designs, this had no impact on piglet survival and performance.

## Supplementation of female piglets from 1 to 28 days of age with a synbiotic: what consequences on cognitive abilities?

*Severine P. Parois[1,2], Morgan B. Garvey[3], Susan D. Eicher[1] and Jeremy N. Marchant-Forde[1]*
*[1]USDA-ARS, Livestock Behavior Research Unit, 270 S Russell Street, 47907 West Lafayette, USA, [2]PEGASE, Agrocampus Ouest, INRA, 16 Le Clos, 35590 Saint-Gilles, France, [3]Purdue University, Animal Sciences, 270 S Russell Street, 47907 West Lafayette, USA; sparois@purdue.edu*

The influence of feed supplements on behavior and memory has been recently studied in livestock species. However, none of the studies in pigs have investigated supplement effects in the same individuals during different cognitive tests. The objectives of the study were to evaluate the effects of a synbiotic on an episodic-like (Object recognition: OR), a working (Barrier solving: BS) and a long-term (T-maze) memory test. A total of 18 female piglets were supplemented from 1 to 28 days of age with a synbiotic (SYN) containing strains of Lactobacillus, fructo-oligosaccharide and beta-glucan included in chocolate milk, whereas a further 17 only received the chocolate milk (CTL). At weaning (means ± SD; age 18.1±1.8 days, weight 13.8±2.4 kg), they were grouped by 8. A period of habituation in the test arena of 3 successive days (2 trials per day of 5 or 10 min) for the piglets to isolation preceded the OR and BS tests. In farrowing crates, they were exposed for 24 h to an object to chew before the OR test to acclimation to objects. The OR test at 16 days of age tested the ability to remember an object already explored 50 min before. The BS test at 20 days consisted of finding a route through two barriers to join two companion piglets, over five successive trials. The T-maze test between 33 and 41 days consisted of finding a food reward. The test was preceded by another 4 days period of habituation to isolation and learning of the food reward. A total of 9 piglets with no interest for the reward were removed from that test. The test was divided into 2 periods: 6 days of acquisition using the same arm rewarded and 3 days of a reversal stage. The treatment effects were evaluated using ANOVA, with repeated measures for the BS test. Treatment groups had no effect on traits of habituation periods ($P>0.1$). In the OR test, both treatment groups explored the reference objects and environment the same way (distance travelled, frequency and duration of interaction, $P>0.1$). SYN piglets interacted quicker with the novel object than the CTL piglets (165±116 vs 255±120 s; $P<0.05$). In the BS test, no differences were found regarding the times needed to cross each barrier and to finish the test ($P>0.1$). Performances in trial 1 were lowest for all traits ($P<0.001$), confirming that piglets used their short-term memory. SYN piglets had shorter distances to finish the test in trial 3 (4.2±1.0 vs 7.4±5.2 m; $P<0.05$). In the T-maze test, number of correct choices and time to succeed were similar ($P>0.1$), except on day 3 of the acquisition stage where SYN piglets were quicker than CTL piglets (13.0±6.2 vs 25.8±18.3 s; $P<0.05$). During the reversal stage, SYN piglets tried the new rewarded arm earlier than CTL piglets (10.3±10.9 vs 19.8±15.4 trials; $P<0.05$). The synbiotic supplement may confer memory advantages in the 3 cognitive tasks, regardless of the nature of the reward and the memory request. However, differences in motivation to solve the tasks cannot be completely excluded.

## Effects of early-socialization and neonatal enriched environment on aggression, performance and weaning ability of piglets

*Heng-Lun Ko[1], Qiai Chong[2], Damian Escribano[1], Xavier Manteca[1] and Pol Llonch[1]*
*[1]Universitat Autònoma de Barcelona, School of Veterinary Science, 08193 Cerdanyola del Vallès, Barcelona, Spain, [2]University of Edinburgh, The Royal (Dick) School of Veterinary Studies, Easter Bush Campus, Midlothian EH25 9RG, United Kingdom; henglun.ko@uab.cat*

The aim of the study was to investigate the combined effect of early-socialization and enriched environment during suckling on performance, social and explorative behaviors, and adaptability at weaning. Forty-eight litters of Danbred piglets were studied from birth to 14D after weaning at an intensive commercial farm of 1,200 sows. Control (CON) litters (n=24) were housed in typical barren farrowing pens, whereas Enriched (ENR) litters (n=24) were provided with 6 enrichment objects (2 hemp ropes, 2 commercial dog rubber chew toys and 2 handmade plastic tube toys) per pen from 1D of age. Sows of both treatment were randomly distributed. Socialization was facilitated by removing the barrier between two adjacent ENR pens at 14D of age. All piglets were weaned and regrouped at 24D of age into 16 barren nursery pens (~38 piglets/pen) according to their body weight but separated by treatment ($n_{CON}$=8; $n_{ENR}$=8). Body weight was recorded on D1, D14, D23, D27, D31 and D38. Live behavioral observation of social and explorative behaviors was conducted twice pre-weaning (D15 and D22) and twice post-weaning (D29 and D36) by scan sampling. Six piglets per litter from seventeen randomly selected litters from each treatment were chosen ($n_{CON}$=102; $n_{ENR}$=102) for assessment of skin lesions and collection of saliva samples. Skin lesions were counted on D14, D15, D23, D25 and D26, while saliva samples were collected on D23, D25 and D26. In pre-weaning stage, social behavior was not significantly different between treatments; ENR piglets spent more time interacting with pen features and enrichment objects ($P<0.001$). Although during early-socialization there were more skin lesions in ENR than in CON ($P<0.001$), the opposite was observed around weaning ($P<0.001$). CON piglets had a significant increase in salivary cortisol ($P<0.001$), chromogranin A ($P<0.001$) and α-amylase ($P<0.01$) during weaning. Average daily gain on 3D after weaning was significantly greater in ENR than CON ($P<0.01$). In conclusion, improving the neonatal environment (combination of early socialization and enrichment objects during suckling) increased explorative (including object play) behavior and enhanced the ability of piglets to adapt at weaning, which leads to better piglet growth.

**Piglet access to chewable materials has a positive effect on sow health and behaviour**
*Kirsi-Marja Swan, Helena Telkänranta, Camilla Munsterhjelm, Olli Peltoniemi and Anna Valros*
*University of Helsinki, Department of Production Animal Medicine, P.O. Box 57, 00014 University*
*of Helsinki, Finland; kirsi.swan@helsinki.fi*

Pigs are known to benefit from access to manipulable materials after weaning. Yet, there is little data on the influence of these on piglet and sow behaviour before weaning. Sows in farrowing crates cannot avoid rooting and chewing directed towards them by the piglets. Teat contacts which do not lead to nursing may be unpleasant and painful for the sow. The aim of the study was to determine if access to chewable materials for piglets during the first weeks of life affected sow health and behaviour. Litters of 59 sows were used in the study. Thirty litters had access to 10 sisal ropes, a plastic ball, newspaper and wood shavings (the Rope-Paper group, RP, n=30). The control group (C, n=29), had access to wood shavings and a plastic ball. The mean litter size was 11 piglets (min 7, max 13). The sows were kept in a standard farrowing crate without access to the materials. Behaviour was recorded for a four-hour period during days 4-18 of life, twice for 51 of the litters. The following behaviours of the sows were analyzed: standing, sitting, sternal recumbency, lateral recumbency, oral-nasal manipulation, eating and drinking. Continuous observation of the four-hour recordings was performed and the frequency and duration of the behaviours were recorded. Mann-Whitney U-tests were used separately for two age groups: 7-13 days old (AGE1) (n(RP)=22, n(C)=22) and 14 days or older (AGE2) (n(RP)=24, n(C)=24). None of the sows were included twice in the same age group. Skin lesions on the sows were observed once weekly, during weeks 1-5 of lactation. The observation was based on the Welfare Quality® protocol, giving each sow a skin lesion score from 0 (<4 lesions) to 2 (>15 lesions). Skin lesion data were analysed with Friedman´s two-way analysis of variance with repeated measures. At AGE 2, sows were standing ($P$=0.01), eating ($P$=0.04) and performing oral-nasal manipulation ($P$<0.01) more often in the C group (median and min-max events of standing 4 (0-11), eating 6 (0-17), oral-nasal manipulation 15(2-31)) than in the RP group (events of standing 2(0-9) eating 2(0-17) and oral-nasal manipulation 5(0-28)). The duration of standing ($P$=0.02) and oral-nasal manipulation ($P$=0.02) was longer in the C group (17(0-68) minutes for standing; 11(1-50) min for oral-nasal manipulation) than in the RP group (6(0-47); 3(0-33)). Skin lesions of the C group differed significantly between the observation weeks ($P$=0.01). Sows tended to have a higher skin lesion score in observation week five with median 2 (1-2) than week four with median 1(1-2) ($P$=0.02). No differences between weeks in skin lesions in the RP group ($P$=0.23) were found with median 1(1-2) on week 4 and 1(1-2) on week 5. The sow manipulated objects, stood up and ate less frequently when piglets had access to chewable materials, which may be due to piglet behaviour towards the sow. The difference in the progression of the skin lesions of the sow may be due to the way the piglets make contact with the sow, which affects the progression of the injuries. Piglet access to chewable materials during early weeks of life may have a positive effect and the use of chewable materials during lactation is thus recommended.

## PigWatch – combining the eye of the stockman and precision farming techniques to improve pig welfare

*Manuela Zebunke[1], Etienne Labyt[2], Christelle Godin[2], Céline Tallet-Saighi[3], Armelle Prunier[3], Sabine Dippel[4], Barbara Früh[5], Glenn Gunner Brink Nielsen[6], Helle Daugaard Larsen[6], Herman Vermeer[7] and Hans Spoolder[7]*
*[1]Leibniz Institute for Farm Animal Biology (FBN), Wilhelm-Stahl-Allee 2, 18196 Dummerstorf, Germany, [2]CEA LETI, Minatec Campus, 38054 Grenoble, France, [3]PEGASE, Agrocampus Ouest, INRA, 35590 Saint-Gilles, France, [4]Friedrich-Loeffler-Institut, Dörnbergstr. 25/27, 29223 Celle, Germany, [5]Research Institute of Organic Agriculture (FiBL), Ackerstrasse 113, 5070 Frick, Switzerland, [6]Danish Meat Research Institute (DMRI), Gregersensvej 9, 2630 Taastrup, Denmark, [7]Wageningen Livestock Research (WLR), De Elst 1, 6700 AH Wageningen, the Netherlands; zebunke@fbn-dummerstorf.de*

Despite decades of intensive research on injurious behaviours like aggression and tail biting in pigs, these problems persist on many farms. In Europe, stricter enforcement of legislation against routine tail docking puts additional pressure on farmers, and researchers are asked to find solutions. However, remedial measures are dependent on early diagnosis of the problem. The European PigWatch Project (https://pigwatch.net, Anihwa ERA-Net) aims to sensitize stock persons to early signs which predict injurious behaviours and to develop automatic measurement techniques that could help farmers to manage their herd. We developed, tested and adapted an on-farm observation protocol in the five participating countries of the project (Netherlands, France, Switzerland, Germany and Denmark) to train stock persons to observe their animals in a different way. The use of the protocol by the farmers showed that the position of the tail (tucked, hanging or curled) seems to be a good indicator for underlying problems that might culminate in harmful behaviour. Some farmers state that using the protocol changed the way they look at their animals. Furthermore, as increased behavioural activity is a clear sign for ongoing aggressive acts, a sensor is being developed in France measuring animal activity and sending an alarm when detecting a sustained fighting. In a first test, the behaviour of 32 pigs (4 pens with 8 pigs each, 110-150 days of age) was observed for 30 h via video and annotated for episodes of fighting (head knock, biting). Three pigs per pen (n=12) were equipped with the activity sensor via an ear tag and alarms were compared with the video analyses. Results showed that the sensors detected 42% of the fighting, with 62% true positives. Its sensitivity and specificity will be improved through additional observations and recordings. Moreover, the sensor will be tested soon on-farm in Germany. In addition to the advantages on the farms, these technologies should enable standardised and efficient animal monitoring. The project will end in 2019, with tools to help farmers to identify early signs of aggression and tail biting in their pigs.

**Is increased aggressive behavior in finisher pigs post-mixing a reliable predictor of lower long-term pen stability?**

*Anna C. Bosgraaf[1], Kaitlin E. Wurtz[2], Carly I. O'Malley[2], Juan P. Steibel[2,3], Ronald O. Bates[2], Catherine W. Ernst[2] and Janice M. Siegford[2]*
[1]*Michigan State University, Department of Integrative Biology, 288 Farm Lane, East Lansing, MI 48824, USA,* [2]*Michigan State University, Department of Animal Science, Anthony Hall, 474 S. Shaw Lane, East Lansing, MI 48824, USA,* [3]*Michigan State University, Department of Fisheries and Wildlife, 480 Wilson Road, East Lansing, MI 48824, USA; bosgraa3@msu.edu*

In modern production, it is common for unfamiliar pigs to be mixed at various stages leading to intense fighting during the first 24-48 h as the pigs establish a social hierarchy. If the hierarchy fails to stabilize, chronic aggression may persist. Previous research has indicated that there may be an inverse relationship between number and type of aggressive interactions immediately post-mixing and number of aggressive interactions observed in a more stable social group. This study further examined this relationship, and investigated whether higher levels of non-damaging aggression (head knocks) immediately after mixing were predictive of lower stability in these groups 3 wk later using similar methods as previous studies. At 10 wk of age, pigs were mixed by weight and sex into finisher pens (10-15 pigs/pen, 12 pens of gilts and 12 pens of barrows). Pens were video recorded for 24 h after mixing and again 3 wk later. Trained personnel recorded durations (in seconds) of damaging aggressive behaviors (reciprocal fighting, attacks, fight withdrawals, and rests during fight) and non-damaging aggressive behaviors (head knocks, bites, inverse and parallel pressing) for 9 h immediately after mixing and 4 h 3 wk after mixing. From this data, total duration of all aggressive behavior was calculated. General linear models were used to compare total duration of damaging aggression in the stable pens with total duration of aggressive behavior at mixing. Models contained fixed effects of head knocking at mixing, sex, number of pigs per pen, average pen weight at mixing, and duration of damaging aggressions post-mixing. No relationship was found between the total duration of damaging aggression in stable groups and head knocking at mixing ($P=0.438$), number of pigs per pen ($P=0.119$), sex ($P=0.434$), weight ($P=0.097$), or total duration of damaging aggression at mixing ($P=0.892$). The total duration of all aggressive interactions in stable groups was also compared with total duration of damaging aggression at mixing, sex, average pen weight at mixing, and number of pigs per pen. No relationships were found between total duration of aggressive behavior in stable groups and total duration of aggressive behavior at mixing ($P=0.279$), sex ($P=0.848$), or weight ($P=0.132$). There was a positive relationship between total duration of aggression in stable groups and number of pigs per pen ($P=0.027$). These preliminary results suggest that in our population there was no inverse relationship between aggression at mixing and aggression in stable groups as was described in previous studies. More research is needed to understand whether differences between this and previous work are due to genetic, environmental or management-based factors, as being able to predict long-term group stability has implications for producers trying to mitigate both acute and chronic aggression.

## How does agonistic behaviour of fattening boars and barrows affect the individual with respect to skin lesions?

*Jeannette Christin Lange and Ute Knierim*
*University of Kassel, Farm Animal Behaviour and Husbandry Section, Nordbahnhofstr. 1a, 37213*
*Witzenhausen, Germany; jlange@uni-kassel.de*

It is well known that young entire boars kept for fattening are more active than castrated ones, and in this connection show more agonistic interactions. However, it is an open question whether this increased agonistic behaviour may be a source of suffering. This would be particularly likely if certain individuals were consistent victims. In an earlier investigation, we observed 43 boar groups and 26 control barrow groups (in 26 batches of castrated and uncastrated siblings) on 5 organic farms and found, despite increased interactions in boars, no differences in lesions or lameness, and no influence of group size (10-25). In the present analysis, we wanted to know whether skin lesions were concentrated on certain individuals and whether this was different between boars and barrows. Therefore, we looked at a sub-set of the data from the same investigated groups including 516 (of 625) ear tagged boars and 272 (of 413) barrows. Skin lesions were scored and counted at one randomly selected bodyside per pig following the Welfare Quality protocol (WQ), and at around a mean group weight of 80 kg liveweight (1st scoring) and just before the first pigs were sent to slaughter (2nd scoring, approx. 34 days later). Weights of 380 boars and 237 barrows were available from the second scoring (mean: 107 kg). Mean lesion numbers were similar in boars and barrows ($5.5\pm4.6$ vs $5.2\pm3.8$ lesions at 1st and $5.0\pm4.2$ vs $4.9\pm3.9$ at 2nd scoring). Only two boars and no barrow received the highest (worst) WQ-score 2; wounds >5 cm were recorded in 1.1% boars and 1.2% barrows over both scorings. In 24.6% of the groups there were pigs with high numbers of lesions (>15) in at least one of the two scorings, of which 3 boar and 1 barrow groups had conspicuously high numbers of affected pigs (up to 41% per group and scoring): 64% of boars with >15 lesions and 55% of barrows belonged to these groups. Overall, the percentages of boars with >15 lesions (3.9% at 1st and 3.3% at 2nd scoring) were not significantly different from those of barrows (1.8 and 2.6%; $\chi^2$test, $P=0.12$ and $P=0.58$). Moreover, only 4 boars and 1 barrow (out of 44 pigs in total) were in this category in both scorings. Outliers in a group with regard to number of lesions, at least if prevalence is higher than usual, (in 8% of boar and 7% of barrow groups), may indicate prolonged disputes between individuals or bullying. Both in boars and barrows, the number of lesions was not associated with liveweight ($P\geq0.12$, linear mixed model in R, fixed factors liveweight, castration status and their interaction, random factors group/farm). Thus, lighter pigs per group were not affected more. In conclusion, we found no significant differences in lesion patterns between boars and barrows, although the consistent numerical differences may reflect a slightly higher welfare risk in boars. Results do indicate a heightened risk for skin lesions in certain groups, but no general concentration on certain victims, although individual cases may occur.

**Comparison of different enrichment materials to fulfil pig's behavioural needs and farmer's management system**
*Emma Fàbrega[1], Roger Vidal[1], Míriam Marcet Rius[2], Alícia Rodríguez[1], Antonio Velarde[1], Damián Escribano[3,4], Joaquin Cerón[3] and Xavier Manteca[4]*
*[1]IRTA, Animal Welfare Subprogram, Veïnat Sies, s/n, 17121 Monells, Spain, [2]IRSEA, Physiological and Behavioural Mechanisms of Adaptation Department, 84400 Apt, France, [3]Universidad de Murcia, Interlab-UMU, 30100 Murcia, Spain, [4]Universitat Autònoma de Barcelona, School of Veterinary Science, 08193 Bellaterra, Spain; emma.fabrega@irta.cat*

Strategies that fulfil pig's behavioural needs for rooting and meet the requirements of intensive management systems are required when raising pigs with entire tails. The objective of this study was to compare four different enrichment materials that are easy to implement on-farm: straw in a rack (S), chains (C), wooden logs (W) and newspaper (N). Four pens per treatment with six entire male fattening pigs each were used (from a mean age of 60 to 165 days). Pigs were observed by scan and focal methods once per week between 9:00 am and 13:00 pm during 12 consecutive weeks (12 scans and 9 min focal sampling/pen/week; 108 min total focal time/pen, 288 min total scan time/pen; total time/pen=6.6 h), to record time budget activities, social interactions and redirected (mouth directed towards flank, tail or ears of another pig) or stereotypic behaviours (repeated pattern without a purpose). Tail (swinging in any direction) and ear (change between position from front to back) movements, and ear movement frequency, were recorded once per week by focal sampling in 8 pens as potential indicators of positive and negative behaviours, respectively (144 min total/pen). Every three weeks body lesions according to Welfare Quality protocol and individual weight were recorded. Two blood samples (before introducing enrichment materials and at end of experiment) and three saliva samples (before treatments, 1.5 months after and at the end) were taken to evaluate the neutrophil:lymphocyte ratio as a marker of hypercortisolism and chromogranin A as a marker of adrenergic system activation, respectively. Chromogranin A was found to significantly decrease at the third sample compared to the first in the straw treatment ($0.64 \pm 0.58$ vs $1.14 \pm 1.46$ µg/ml, $P<0.05$), whereas the decrease for the other three treatments was not significant. An increase of frequency of stereotypic behaviours was found for the N and C treatments compared to S and W (3, 1.5 vs 0.01% time spent, respectively, $P<0.1$), whereas redirected behaviours tended to decrease for the S compared to N treatment (13 vs 31% time spent, respectively, $P<0.1$). A higher frequency of tail movements was found when pigs were interacting with the enrichment material compared to when they were not (52 vs 45%, $P<0.05$). These results suggest benefits for animal welfare when providing straw in a rack, although quantities of straw which impaired the slurry management system were found at the end of the experiment inside the pit. Therefore, more work is recommended to develop appropriate strategies for the provision of straw.

## Social preferences among Přeštice black-pied pigs

*Sébastien Goumon¹, Gudrun Illmann¹, Iva Leszkowova¹, Anne Dostálová¹ and Mauricio Cantor²*
*¹Institute of Animal Science, Ethology, Přátelství 815, 104 00, Prague, Czech Republic,*
*²Universidade Federal de Santa Catarina, Ecology and Zoology, Centro de Ciências Biológicas,*
*Campus Universitário, Trindade Caixa Postal 5102, 88040-970 Florianópolis, Brazil;*
*sebastien.goumon@gmail.com*

Social preferences beyond parent-offspring bonds have been largely understudied in social farm animals. Pigs may commonly experience disruptions of their social bonds when mixed at weaning and during the growing-fattening period. Thus, identifying and maintaining pairs of preferred partners may be a way of ensuring better welfare in pigs. The aim of this study was to investigate whether pigs in a stable social group formed non-random associations, and whether the latter were influenced by gender, relatedness and social dominance. Spatial proximity (≤30 cm) was used as a proxy of social association between individuals in two behavioral contexts (active: foraging, exploring, fighting and wallowing; inactive: lying down). Proximity data were collected over 33 h of observations (age 17 to 29 weeks) in a group of 24 Přeštice black-pied pigs (3 litters of males and females of the same age) kept together from 12 weeks of age and housed in an indoor pen with free access to a large outside pastured enclosure (50×60 m). Association indices (simple ratio index, SRI) were calculated for each pair based on proximity data over daily sampling periods, and permutations of the data stream were carried out to test the null hypothesis of random associations against the alternative that pairs of individuals have avoided or preferred associations. Associations were defined as preferred when SRI>0.12 (mean SRI for active and inactive dataset combined) and avoidance when SRI≤0.12. The influence of individual covariates on the proximity-based associations was tested using multiple regression quadratic assignment procedure (MRQAP). Patterns of associations were found to be non-random between sampling periods and dependent on the behavioral context. Non-random associations were less evident during activity (observed SRI coefficient of variation=1.403, random CV=1.156, $P=0.01$, 7.6% of all pairs were potentially preferred) than during inactivity (observed SRI CV=0.727, random CV=0.470, $P<0.001$, 36.6% of potential preferences). Gender tended to have a minor effect on the formation of those associations (MRQAP partial correlation = 0.115, $P=0.054$), while relatedness and dominance had no significant influence on the formation of dyadic associations ($P=0.984$ and $P=0.742$, respectively). In conclusion, the results of this study show that domestic pigs form non-random associations, which seem to be mainly determined by factors other than gender, relatedness and dominance. This study also highlights the importance of considering behavioral state when examining individual relationships as the relevance and benefits of the associations may vary according to the behavioral context. Additional research will focus on other predictors of association such as personality and will include multiple social interactions as more refined proxies to test for social preferences and affiliations.

## Improving our understanding of assessing fear in sheep

*Paul H. Hemsworth, Maxine Rice, Ellen C. Jongman, Lauren M. Hemsworth and Grahame J. Coleman*
*University of Melbourne, Animal Welfare Science Centre, Faculty of Veterinary and Agricultural Science, Parkville, 3052, Victoria, Australia; phh@unimelb.edu.au*

Fear has important implications on the welfare of commercial sheep. Many tests have been utilised to assess animal fear, however different emotional experiences and motivations are involved in the perception of and reaction to the stimuli in these tests. Our understanding of the behavioural indicators of these emotional experiences and motivations is poor. Eighty-seven weaned lambs (average live weight of 43.3 kg) were individually exposed to three commonly-used tests involving novelty, restraint, human contact and social isolation: a 1-minute open field test (OFT); continuous approach by a human in a 1-minute flight distance test (FDT); and close restraint in a 2-minute isolation box test (IBT). The behavioural variables measured in the tests were subjected to a principle component analysis (PCA) with a varimax rotation to identify sets of components that represent the underlying commonalities (components) in these tests. Component scores were calculated for each component as the sum of the standardised item scores multiplied by their corresponding coefficient for that component. An inspection of the correlation matrix for the PCA revealed that the correlation coefficients were all above the required 0.3. The Kaiser-Meyer-Olkin value exceeded the recommended value of 0.6, and Bartlett's Test of Sphericity reached statistical significance. Thus, the factorability of the each of the correlation matrices was supported. The PCA extracted three components, all with eigenvalues greater than 1. The first component, which was labelled 'Bleats' and loaded heavily and positively on the number of vocalizations in the three tests, accounted for 17% of the variance. The second component, which was labelled 'Human avoidance' and loaded heavily and positively on the flight distance at 15 s time points in the FDT, and moderately and negatively on numbers of jumps in the IBT and escape attempts in the OFT, accounted for 14% of the variance. The third component, which was labelled 'Activity' and loaded heavily and negatively on steps and turns in the IBT and moderately and negatively on squares entered in the OFT and time taken to move to the testing area, accounted for 11% of the variance. These results indicate that the three tests (or combinations thereof) may reflect fear in different aversive situations, such as fear of social isolation, close presence of humans and novelty. Vocalizations in the three tests may be predominantly a response to social separation. Activity in the OFT and IBT and resistance to moving to the testing area may be predominantly a response to novelty, while avoidance of the human in the FDT is likely to be a fear response to humans. The combined use of these tests appears well suited to study stimulus-specific fear responses in sheep, but the validity of the measurements need to be tested, for example through research in which motivations for social reinstatement and avoidance of humans and novelty is each experimentally manipulated.

**Effects of flooring surface and a supplemental heat source on dairy goat kids: growth, lying times and preferred location**

*Mhairi Sutherland, Gemma Lowe and Karin Schütz*
*AgResearch Ltd., Ruakura Research Centre, Hamilton 3214, New Zealand;*
*mhairi.sutherland@agresearch.co.nz*

Dairy goat kids are typically born in the middle of winter in New Zealand, separated from their dams at birth, and reared on surfaces such as wood shavings or metal mesh. The objective of this study was to evaluate the effects of different flooring surfaces and a supplemental heat source on lying behaviour, preferred lying location and growth rates in kids. Eighty female Saanen dairy goat kids were enrolled in the study at 3±0.9 d of age (mean ± SD) and allocated to one of four treatment pens (n=4 pens/treatment, 5 goats/pen): (1) wood shavings with two 250 W heat lamps (WS+H); (2) wood shavings without heat lamps (WS); (3) metal mesh with two heat lamps (MM+H); or (4) metal mesh without heat lamps (MM). Kids were reared in treatment pens (1.5×3.5 m) for 8 days. Lying behaviour was recorded continuously using accelerometers (lying times and bout information). Each pen was divided into three equally sized zones (warm, moderate, cold) and time spent lying in each zone was determined from video recordings. The average temperature (mean ± SD) in the different zones in pens with heat lamps was: warm: 18±0.7 °C, moderate: 15±0.4 °C, cold: 14±0.2 °C, and in pens without heat lamps the average temperature in the corresponding zones was: warm: 14±0.3 °C, moderate: 14±0.2 °C, cold: 14±0.2 °C. Milk consumption was recorded daily and body weight was recorded at the start and end of the trial period and body weight gain calculated and summarised on a group level. Data were analysed by ANOVA and results are presented as least square means ± SED. Kids managed on wood shavings (with and without heat lamps) spent approximately 2 h more lying per day than kids housed on metal mesh (with and without heat lamps). Kids housed on wood shavings with heat lamps spent more time lying than all other treatments (lying time, h/24 h: WS+H: 18.1±0.38; WS: 16.2±0.38; MM+H: 15.0±0.38; MM: 15.1±0.38; $P=0.005$). Lying times were similar between kids reared on metal mesh with or without heat lamps. Kids managed on wood shavings with heat lamps performed more lying bouts than all other treatments (no./24 h: WS+H: 24±0.8; WS: 19±0.8; MM+H: 18±0.8; MM: 19±0.8; $P<0.001$). Kids preferred to lie in the warm zone when a supplemental heat source was provided regardless of flooring type (WS+H: 98±7.4; MM+H: 94±7.4% of total lying time in the warm zone; $P<0.001$). Milk consumption was not affected by treatment, however, kids managed on wood shavings tended ($P=0.052$) to gain more weight than kids managed on metal mesh (1.05±0.059 and 0.91±0.059 kg, respectively, over the 8 d of recording). Provision of a supplemental heat source had no effect on weight gain. Kids had a clear preference for lying in the warm zone of the pen when a heat lamp was provided. Longer lying times and greater weight gain suggest that rearing kids on wood shavings improves kid comfort and performance.

## The effect of outdoor space provision on lying behaviour in lactating dairy goats

*Hannah B.R. Freeman[1,2], Jim Webster[2], Gemma L. Charlton[3], Klaus Lehnert[1] and Gosia Zobel[2]*
*[1]The University of Auckland, School of Biological Sciences, Auckland, 1142, New Zealand,*
*[2]AgResearch Ltd., Ruakura Research Centre, 10 Bisley Road, Private Bag 3123, Hamilton 3214,*
*New Zealand, [3]Harper Adams University, Animal Production, Welfare and Veterinary Sciences,*
*Shropshire, TF10 8NB, United Kingdom; hannah.freeman@agresearch.co.nz*

Time spent being active is important for the welfare of ruminant animals. Lying time is an indicator of general levels of inactivity. One technique for increasing activity in housed commercial animals is providing outdoor access. The aim of this study was to examine the effect of providing two types of outdoor space (empty and enriched) on lying behaviour of dairy goats in a New Zealand commercial system. We hypothesised that outdoor access would reduce lying time, and that this would be more pronounced when outdoor space was enriched. Two-hundred and eighty-eight lactating Saanen X does (4 groups of 36 goats, over 2 consecutive reps) were studied, between Nov. 2017 and Jan. 2018. During the baseline phase, each group of goats were housed indoors in a 10×12 m pen with a surface of 30% concrete and 70% wood shavings (approx. 30 cm deep). This was followed by two treatment phases of free-access to either empty or enriched outdoor space (sand-pack, 16.5×7 m), in a crossover design. Enriched space contained additional items designed to encourage natural behaviours: climbing platforms, shelters, scratching brushes, manipulable abacuses, and paved areas. Data loggers were used to record lying behaviour of 10 goats/pen for 72 h per phase, and a daily average was calculated for each goat. PROC GLIMMIX (SAS) models were used to assess the effect of outdoor housing type on lying time and lying bout frequency, with rep as a fixed effect, and test within group and goat within group as random effects. All results are given as mean ± SED, or mean with 95% confidence interval [lower limit, upper limit] as applicable. During the baseline phase (indoor-access only) goats spent 14.6±0.3 h/d lying. This decreased by 1 h/d ($P<0.0001$) when goats had free-access to outdoor space, irrespective of whether this space was enriched or empty ($P>0.1$). There was no difference in the number of lying bouts between baseline and treatment phases, 23 [21, 25] bouts/d ($P=0.8$). Rep had an effect on the model. Overall, goats in rep 1 lay for 1.8 h less per day ($P=0.003$) and had 6 [5, 8] bouts/d fewer ($P<0.0001$), compared to goats in rep 2. This could be due to an increase in temperature as the summer progressed (1.7 °C increase in mean max. daily temperature in rep 2). Provision of outdoor space decreased lying time. The benefits of this have not yet been explored in goats, however, in cows increased activity has been linked to improved foot and leg health, and reduced somatic cell count. The effect of outdoor access on lying time was not altered by providing enrichments, however, enrichments may have changed other aspects of behaviour, which have not yet been analysed.

## Positive handling of goats does not affect human-directed behaviour in the unsolvable task

*Jan Langbein, Annika Krause and Christian Nawroth*
*Leibniz Institute for Farm Animal Biology, Institute of Behavioural Physiology, Wilhelm-Stahl-Allee 2, Haus 7, 18196 Dummerstorf, Germany; langbein@fbn-dummerstorf.de*

Next to domestication, the degree of socialization with humans has been proposed to impact on the understanding of human social cues by domestic animals. A test paradigm that is used to demonstrate human-animal interaction is the 'unsolvable task'. Subjects are confronted with a task that is first solvable during training but then unsolvable in the test while they have the possibility to interact with a human experimenter. We investigated human-directed behaviour in the unsolvable task by goats which did not receive human handling other than management routines as well as the impact of sustained positive handling on these behaviours. The study was conducted with two groups of 10 goats each (all female, 10-16 months old). Before the experiment, group 1 (G1) was positively handled (including stroking and feeding) in their home pen by an experimenter for 30 minutes twice-daily over a period of two weeks, while group 2 (G2) was not. Next, both groups were trained to remove a lid from a clear box to receive a reward. G1 received human demonstration on how to remove the lid, but in G2 the experimenter only repeatedly baited the box, put the lid on, and left the pen immediately. The respective procedure was repeated two times a day over five days. Afterwards, both groups received 20 solvable training trials to open the box in a test arena with the experimenter sitting beside the box. Training trials were completed on four consecutive days. Finally, in three nearly identical test trials (unsolvable) the lid was fixed to the box. We measured time of contact with the box as well as number of gaze alternations/contact alternations with the experimenter plus the respective latency to first gaze or contact alternation in the three test trials. All goats reached the training criterion of six successful trials in a row within the 20 training trials with no difference between groups (U=50.5, $P$=0.53). In test trials, both groups did not differ in their time of contact with the box (F1,18=0.16, $P$=0.691), while time of contact decreased over repetitions (F2,18=73.99, $P$<0.001). The number of goats that gaze-alternated at the experimenter in the three test trials was 7, 9 and 10 in G1 and 8, 10 and 9 in G2 with no impact of group (F1,17.9=2.48, $P$=0.133) or repetition (F2,17.7=2.63, $P$=0.099). The number of goats making contact alternation with the experimenter was 3, 7 and 7 in G1 and 5, 6 and 4 in G2, again, with no impact of group (F1,16.0=0.08, $P$=0.786) or repetition (F2,17.0=0.87, $P$=0.437). However, latency to either first gaze alternation or first contact alternation decreased over the three test trials (F2,15.4=42.057, $P$<0.001; F2,10.1=32.925, $P$<0.001, respectively). In conclusion, while the latency of the goats to gaze or contact alternations decreased, their motivation to open the box decreased. This might indicate that human-directed behaviour occurs once subjects experience that they cannot solve the problem themselves. We found that goats show human-directed behaviour in the unsolvable task without prior positive handling. Thus, domestication may have resulted in goats being predisposed to interacting with humans, as are dogs.

**Evaluation of alternatives to cautery disbudding of dairy goat kids using behavioural measures of post-operative pain**

*Melissa N. Hempstead[1,2], Joseph R. Waas[1], Mairi Stewart[3], Vanessa M. Cave[2] and Mhairi A. Sutherland[2]*
[1]*The University of Waikato, The School of Science, Private Bag 3105, 3240 Hamilton, New Zealand, [2]AgResearch Ltd., Ruakura Research Centre, 3214 Hamilton, New Zealand, [3]Interag, Ruakura Research Centre, 3214 Hamilton, New Zealand; melissa.hempstead@agresearch.co.nz*

Cautery disbudding is a painful husbandry practice that can negatively impact goat welfare. Clove oil, which has anaesthetic and cytotoxic properties is a novel disbudding technique. Cryosurgical, caustic paste and clove oil disbudding may cause less pain than cautery disbudding in goat kids. We evaluated alternatives to cautery disbudding (i.e. caustic paste, cryosurgery and clove oil) for goat kids using behavioural measures of post-operative pain. Fifty Saanen doe kids aged 10.6±0.91 d (mean ± SD) were randomly assigned to one of five treatments (n=10/treatment): cautery disbudded using a hot cautery iron (CAUT), caustic paste disbudded using a sodium hydroxide-based paste rubbed into the horn buds (CASP), cryosurgical disbudding using a pressurised spray of liquid nitrogen onto the horn buds (CRYO), clove oil injected (0.2 ml) into the horn buds (CLOV) or sham disbudding using a finger to massage the horn buds (SHAM). Head and body shaking, head scratching, self-grooming and feeding behaviour were recorded for 1 h pre- and post-treatment. Total frequencies of each behaviour were measured, as were the total durations of head scratching, self-grooming and feeding (mean ± SED). Accelerometers measured individual lying bouts, bout duration and total lying time over 24 h pre- and post-treatment. Data were analysed using a one-way analysis of covariance (blocked by treatment date and pen) adjusted for the pre-treatment values. CASP kids performed more head shakes and head scratches, but less self-grooming and lower feeding durations than CAUT kids (73.3 vs 38.5±11.06 head shakes, 35.1 vs 13.1±6.62 head scratches, 1.3 vs 8.7±2.00 self-grooms and 1.0 vs 2.4 min/h; $P \leq 0.05$). CRYO kids also performed higher frequencies of head shaking (55.9±11.06 head shakes) and head scratching (28.8±6.62 head scratches) than CAUT kids ($P \leq 0.05$). Head shaking, head scratching and self-grooming frequencies in CLOV kids were no different to those of SHAM kids (34.0 vs 38.1±11.06 head shakes, 16.7 vs 10.5±6.62 head scratches and 12.6 vs 1.3±2.00 self-grooms; $P > 0.10$). There was no difference in the number of lying bouts for SHAM kids compared with CAUT, CLOV and CRYO kids (29.9, 26.3, 33.4, and 32.8±2.25 no. bouts; $P > 0.50$). CAUT kids displayed fewer lying bouts than CLOV and CRYO kids (33.4 and 32.8±2.25 no. bouts, $P \leq 0.01$); however, CAUT kids spent more time lying than SHAM, CRYO and CLOV kids over 24 h post-treatment (17.0, 16.2, 15.8 and 16.1±0.32 h/24 h; $P \leq 0.05$). There was no effect of treatment on lying bout duration over 24 h post-treatment ($P=0.09$). It appears that caustic paste and cryosurgical disbudding were more painful than cautery disbudding and may not be suitable alternatives for goat kids. Clove oil injection shows promise as an alternative to cautery disbudding; however, future research should evaluate the long-term effect of clove oil on behaviour and its efficacy in preventing horn growth.

### Changes in lying time after hoof trimming in dairy goats

*Laura Deeming[1,2], Heather Neave[3], Ngaio Beausoleil[1], Kevin Stafford[1], Jim Webster[2] and Gosia Zobel[2]*
*[1]Massey University, School of Veterinary Science, Palmerston North, 4410, New Zealand, [2]AgResearch, Animal Welfare, Ruakura Research Centre, 10 Bisley Road, Hamilton 3214, New Zealand, [3]University of British Columbia, Animal Welfare Program, Vancouver, British Columbia, Canada; laura.deeming@agresearch.co.nz*

Routine hoof trimming improves gait score, reduces hoof lesions and lameness and optimises weight distribution between the two claws. However, the process of hoof trimming is associated with immediate behavioural effects in dairy cows, such as a change in lying behaviour. To date there are no data investigating the impact of hoof trimming in dairy goats. Therefore, the aim of this study was to describe daily lying time before and after trimming in dairy does. Sixty-four Saanen cross does were enrolled on a commercial dairy goat farm in Waikato, New Zealand. The goats were housed in a pen bedded with shavings and a concrete skirt in front of the feed rail. The goats' hooves were trimmed at 18 (just prior to second mating), 22 (while in kid), and 26 months of age (in early lactation). Prior to trimming, each goat was fitted with a data logger to measure lying time from 3 d before to 3 d after hoof trimming. Daily lying time (h/d) was averaged for the 3 days before trimming (baseline) and compared to lying time on each of the 3 days after trimming (d1, d2, d3) for each trimming session (18, 22 and 26 mo). A mixed model analysis (PROC MIXED, SAS) was conducted for assessing the effect of trimming and session on daily lying time with day in relation to trimming within trimming session as repeated measures. Over all sessions, lying time was greater on each of the 3 days after trimming compared to before trimming (mean ± SED h/d: baseline = 12.8 h/d, $P<0.01$; d1=14.5±0.13 h/d; d2=13.3±0.14 h/d; d3=13.2±0.13 h/d,). The greatest increase in lying time occurred on d1 (d1 vs d2: $P<0.001$; d1 vs d3: $P<0.001$). There was no significant difference in lying time between d2 and d3. Over all days, lying time was greater when the goats were in kid (13.9±0.13 h/d) compared to at mating (13.1±0.13 $P<0.001$) and in early lactation (13.1±0.13 $P<0.001$). In dairy cows, an increase in lying time after trimming is associated with a short term increase in lameness scores, which suggest potential pain. Therefore, in the current study, the increase in daily lying time in the 3 days following hoof trimming may be due to goats experiencing pain or hoof tenderness as a result of trimming. Alternatively, it could be a compensatory response due to the goats being deprived of approximately 4 hours lying time during the trimming process. The greater daily lying time at 22 months may be due to the goats being in kid, or it could be due to them not being involved in twice daily milking. These results show that hoof trimming has an initial effect on lying behaviour in dairy goats, and that timing of trimming during lactation and gestation may impact the goat's behavioural response. Future research could determine the reasons for the increased lying time in dairy goats after hoof trimming and the implications for the goats.

## Behaviour of dairy goats managed in a natural alpine environment

*Gosia Zobel[1], Hannah Freeman[1,2], Derek Schneider[3], Harold Henderson[1], Peter Johnstone[1] and Jim Webster[1]*
[1]*AgResearch Ltd., Animal Welfare Team, Ruakura Research Centre, Hamilton 3214, New Zealand,* [2]*University of Auckland, School of Biological Sciences, Auckland 1142, New Zealand,* [3]*University of New England, School of Science and Technology, Armidale 2351, Australia; gosia.zobel@agresearch.co.nz*

Dairy goats are often housed indoors. To inform housing improvements that may benefit the welfare of commercial goats, we need to understand their natural behaviour when they are given a choice of environment. The aim of this study was to first describe behaviours of milking goats in a natural environment and second to determine what factors may impact these behaviours. In July 2017, a herd of approximately 100 free-ranging goats in the Ticino region, Switzerland were monitored from their home hut (1,700 m a.s.l., 8°46'3"E 46°15'7"N). The herd left the hut in the evenings and split into two separate groups, so 10 late lactation does (mixed dairy breeds, mean ± SD: 63±8 kg weight; 2.3±1.0 kg/d milk) from each group were selected for focal observation. All focal goats were fitted with coloured IDs and GPS collars, and half with accelerometers. Eleven d of live observation, accelerometer, and temperature data, and 8 d of GPS data, were recorded. Time and group effects on live observations were tested with mixed models, and results are presented as mean ± SED. The remaining results are presented as mean ± SD. In the evenings, one group climbed NW slopes and the other climbed SE slopes. GPS data indicated that goats traveled 3.3±1.5 km (range: 0.3-9.0 km) each night, with potential elevation changes of 305±132 m (range: 19-610 m). Goats travelled (mean speed: 0.1 m/s) along ridgelines, and camped in areas slightly below these. During the morning 2 h observation period (starting 09:24 h ± 73 min), goats spent 49±3% of their time lying, with 44±4% of this being on exposed rocky surfaces. During the afternoon 2 h observation period (14:11 h ± 17 min), a similar pattern in lying was seen, however the majority was under shade structures (26±3%) or in caves (49±3%) which had a constant temperature (11±0.5 °C). Though the 2 groups travelled in different directions, both spent the majority of the 3 h evening observation period (17:15 h ± 60 min) on sloped ground, either walking (30±1%) or browsing (35±2%). Compared to morning and afternoon, the evening observation period had more browsing (3 vs 35±2% of observed time, $P<0.001$) and grazing (1 vs 11±1%, $P<0.001$). Overall, goats in NW group were observed ruminating less than those in SE group (29 vs 41±3%, $P=0.001$), however, period had an effect (35 vs 55 vs 14±3%, $P<0.001$ for morning, afternoon, evening, respectively) for both groups. Accelerometers indicated that goats spent the majority of their time (47±6.2 min/h) between 23:00 h and 06:00 h lying. A second less pronounced (32±8.0 min/h) lying peak was noted between 10:00 h and 16:00 h. Temperature did not play a significant role in this pattern. Distinct behavioural patterns occurred based on location and time of day. These results suggest that when milking goats have access to environmental diversity, it promotes a variety of behaviours, including lying in the sun, seeking shade, browsing vegetation, climbing and walking extended distances.

**Using qualitative behaviour assessment and flight distance to examine the farmer-ewe relationship in extensive systems**

*Carolina A. Munoz, Paul H. Hemsworth, Grahame J. Coleman, Lida Abrie and Rebecca E. Doyle*
*Animal Welfare Science Centre, The University of Melbourne, 21 Bedford street, North Melbourne,*
*Victoria 3051, Australia; cmunoz@student.unimelb.edu.au*

The quality of the human-animal relationship (HAR) is a key factor affecting animal behaviour and welfare. However, methods for practically assessing the HAR in sheep are limited. This study investigated the use of qualitative behaviour assessment (QBA) and flight distance (FD) as tools to examine the HAR in extensively managed ewes. It was hypothesized that ewes will be more reactive (e.g. alert and agitated) towards human stimulus and will have greater FD during a standardized FD test if they are managed by a farmer with negative attitudes towards sheep and sheep handling. Thirty-two sheep farms in Victoria, Australia were involved in this study, and visits were conducted after weaning (Oct-Dec 2016). Farmers answered an attitude questionnaire and ewes went through a QBA and FD test. In the questionnaire, farmers rated a set of statements on their attitudes towards sheep and sheep handling. To conduct the QBA and the FD test, at each farm 100 ewes were allocated in four groups of 23-27 ewes. The test was performed using a holding pen within the farms' regularly-used sheep yards. A single unfamiliar human stimulus (female), quietly entered the pen (holding one of the four groups of ewes), and walked around the perimeter at 1-step per second. During this procedure, the behaviour of the group was continuously recorded for 1 min by a GoPro camera. After this, FD was measured in 5 ewes randomly selected from each of the four groups of ewes (n=20 ewes tested per farm). The test ended whenever the ewe stepped away from the experimenter, and the FD was estimated. Using the video footage, one observer (not involved in the on-farm data collection) conducted the QBA. The attitude questionnaire and QBA data were each analysed by principal component analysis (PCA), and Spearman rank correlations were performed between the PCA data and the FD data to examine relationships. PCA distinguished three main dimensions of farmer attitudes: PC1 (explained 26% of the variation) summarised as 'positive attitudes towards sheep', PC2 (24%) 'positive attitudes towards sheep handling' and PC3 (14%) 'positive perceived control of sheep handling'. Regarding QBA, one dimension of sheep expression was found: PC1 (77%) ranging from vigorous/agitated/fearful/alert/aggressive to calm/relax/quiet, summarised as 'mood'. Overall, FD ranged from 1.6 to 6.8 m (SD±1.1), and a significant positive correlation was obtained between 'mood' and FD ($r=0.49$ $P=0.004$), indicating that ewes that are more reactive towards human stimulus (vigorous/agitated/fearful/alert/aggressive) also presented greater FD. However, no relationships were found between 'mood' and farmers attitudes or between FD and farmer attitudes. These results suggest a moderate relationship between 'mood' and FD, and no clear relationships between 'mood' or FD and farmer attitudes towards sheep and sheep handling. Further research on farmer attitudes and behaviour specifically during routine husbandry is required to understand the farmer-ewe relationship in extensive systems.

## Effects of enrichment and social status on enrichment use, aggression and stress response of sows housed in ESF pens

*Jennifer A. Brown[1], Cyril R. Roy[1], Yolande M. Seddon[2] and Laurie M. Connor[3]*
*[1]Prairie Swine Centre, Ethology, Box 2105, S7H 5N9, Saskatoon, SK, Canada, [2]University of Saskatchewan, WCVM, College Ave, S7N 5C9, Saskatoon, SK, Canada, [3]University of Manitoba, Animal Science, 66 Chancellors Cir., R3T 2N2, Winnipeg, MB, Canada; jennifer.brown@usask.ca*

Canadian sow herds are transitioning to group housing, and most sows will be housed on partially or fully slatted floors. Many producers are reluctant to provide straw or fibrous materials, thus, it is important to identify what object enrichments can be used to maximize enrichment utilization and welfare benefits in sows. This study observed the effects of four enrichment treatments in dominant (D) and subordinate (S) sows. Five static groups of 20±2 sows were studied from weeks 6 to 14 of gestation in partially slatted pens with electronic sow feeders (2.5 m$^2$/sow). Six focal animals (3 D and 3 S) were selected per group in week 6 of gestation using a feed competition test. Each group received four treatments in random order, with each treatment lasting for 12 days followed by a 2-day break. Treatments consisted of: (1) constant provision of wood on chains (Constant); (2) rotation of objects 3×/week (rope, straw, wood on chain: Rotate); (3) rotation of objects 3×/week with an associative stimulus (bell or whistle; Stimulus); and (4) no objects (Control). Focal sow results are presented. Enrichment interaction was determined using digital photos collected at 15-min intervals for 8 h on days 1, 8, 10 and 12. Sow posture and behaviour were recorded in live observations at 15-min intervals on days 2 and 11 (09:30-10:30, and 01:30-02:30 h). Skin lesion scores were assessed using a 4 point scale on days 1 and 12, and salivary cortisol samples were collected in weeks 6, 10 and 14. Effects of social status, treatment, day and their interactions were analyzed using Proc Mixed and Proc Glimmix in SAS 9.4. Sows spent more time contacting and near enrichments in Rotate and Stimulus treatments than in Constant (Contact duration [% of observations], LSMeans ± SEM, Rotate: 38.21; Stimulus: 37.83; Constant: 18.68±1.95. $P<0.05$). Subordinate sows spent more time in contact with and <1 m from enrichments than D sows (Contact duration [% of observations], LSMeans ± SEM, D: 29.61; S: 33.55±1.95. $P=0.04$). Subordinate sows were observed standing more and lying less than D sows, and D sows were observed object biting more than S sows. On d 1, sows in the Stimulus treatment had higher lesion scores than those in Constant, with Rotate and Control intermediate. Subordinate sows received more skin lesions on day 1 (Shoulder lesion score LSMeans, D: 0.66; S: 1.71; SEM ±0.12. $P<0.01$), and had higher salivary cortisol concentrations than D sows (Cortisol LSMeans, D: 0.24; S: 0.43 μg/dl; SEM ±0.04. $P=0.001$). In conclusion, Rotate and Stimulus treatments resulted in greater sow interaction with enrichment, but Constant enrichment resulted in less aggression. Subordinate sows were more active and spent more time contacting enrichment, but received more aggression and had higher cortisol concentrations than D sows. The results suggest that when access to enrichment is limited and is valued by sows, it can result in greater aggression directed towards subordinates.

## Motivated for movement – a comparison of motivation for exercise and food in stall-housed sows and gilts

*Mariia Tokareva[1], Jennifer A. Brown[2], Alexa Woodward[1], Edmond A. Pajor[3] and Yolande M. Seddon[1]*
[1]*University of Saskatchewan, 52 Campus Drive, S7N 5B4, Saskatoon, SK, Canada,* [2]*Prairie Swine Centre Inc., Box 21057, 2105 8th Street East, S7H 5N9, Saskatoon, SK, Canada,* [3]*University of Calgary, 3330 Hospital Drive NW, T2N 4N1, Calgary, AB, Canada; mariia.tokareva@usask.ca*

The Canadian Code of Practice for the Care and Handling of Pigs permits the operation of existing stall barns after July 2024, provided that mated gilts and sows are given opportunities for greater freedom of movement. Whether stall-housed sows are motivated for movement, and how this is influenced by their prior stall experience (PSE), has not been investigated. The objectives of this study were to determine the motivation of stall-housed female pigs for access to an alleyway for three minutes (M: movement), and compare this to their motivation to receive a portion of the daily feed ration (F: 30% of daily ration; with the other 70% fed to animals 2 hrs prior to testing), and to also assess the influence of animal PSE on motivation for each reward. Stall-housed gilts (n=12) and sows (n=12, parity 2-4) were studied in six replicates (2 sows and 2 gilts/rep), starting one week post breeding. Animals were trained to use an operant panel with two buttons: (1) active button (AB: push counts resulted in a reward of feed or access to the alley); (2) dummy button (DB: push counts not rewarded, to monitor that the sow made a defined choice), and to perform up to nine consecutive AB presses over three days, after which animals were tested on an ascending fixed ratio (FR), with the required number of AB presses increased by 50% each consecutive day. Upon reaching their individual maximal FR for one reward, animals were retrained and tested on the opposite button and received the alternate reward, so the sow was her own control. To monitor the effect of prior training experience (PTE), whether animals were first trained and tested for exercise or feed was swapped per replicate. Animals were given 30 minutes each day to reach the required FR and access the reward. Data were analysed by GLM (SAS 9.4) to determine the influence of reward type and PSE on the highest price paid (HPP, maximum AB push counts in one session) and latency to make the first AB push, in the session that the HPP was reached. There was an interaction between PSE and type of reward on HPP. Sows showed a greater HPP for feed than movement, but for gilts it did not differ. Sows had a greater motivation to access feed than gilts. However, gilts and sows did not differ in their motivation to access movement (mean HPP ± SEM for sows F: 369.25±56.47 vs M: 68.5±13.61; gilts F: 211.67±47.73 vs M: 77.75±19.47, F=4.23, P<0.05). PTE did not influence the HPP for feed or movement by sows or gilts. The latency to press the AB was not influenced by PSE or reward. Results suggest that stall-housed gestating sows and gilts are motivated to access time out of their stall. The levels of motivation for both rewards are equal in gilts, but in sows the motivation for movement is moderate when compared to their greater motivation for feed. The greater motivation to receive a feed reward in sows may be because they were recovering from lactation during the testing period.

## Effects of (a switch in) enriched vs barren housing on the response to reward loss in pigs in a negative contrast test

*Lu Luo, Inonge Reimert, Sharine Smeets, Elske De Haas, Henk Parmentier, Bas Kemp and J. Elizabeth Bolhuis*

*Wageningen University & Research, Animal Sciences, P.O. Box 338, 6700 AH Wageningen, the Netherlands; lu.luo@wur.nl*

Several studies suggest that animals in a negative emotional state are more sensitive to reward losses as shown by behavioural and neurophysiological responses. In a successive negative contrast (SNC) test, reward losses are induced by decreasing the size of the reward for a task for which animals have been trained. This SNC paradigm has not been widely used in pigs. It is well known that environmental enrichment positively influences the welfare of pigs, and may induce a more optimistic emotional state, which could reduce their sensitivity to reward losses. We studied pigs in barren (B) or enriched (E) housing, experiencing either a switch in housing conditions at 7 weeks of age or not (4 treatment groups: EE, EB, BE, BB, n=8 pens per group) in an SNC runway task. We hypothesized that B housed pigs, particularly those that changed from E to B housing, would show an enhanced sensitivity to reward losses. One pig per pen was trained to run a 24.6 m U-shaped runway for 6 pieces and one for 1 piece of apple. Each pig received 3 trials per day, with a maximum of 120 sec/trial. Latency to start eating the reward was recorded, and the average was calculated per day. After 11 days, all pigs received 1 piece of apple only for another 11 days (reward shift: 6-1 vs 1-1 reward group), i.e. the group originally receiving 6 pieces of apple experienced a reward loss. Effects of pre-housing, post-housing, (original) reward size, day and interactions were analysed using mixed models with a random effect of animal. Fifty-one pigs were successfully trained. Before the reward shift, over the first 11 days, pre-housing × post-housing × reward size affected the average run-time ($P<0.05$). All BB pigs ran slower than other pigs (BB: 59.3±2.8; BE: 35.9±1.7; EB: 39.6±2.2; EE: 40.9±2.2, $P<0.05$), without any other significant pairwise differences. Analysis per treatment revealed, however, that EB 6-reward pigs were faster than the 1-reward pigs. Overall latency was higher on the last days ($P<0.001$). After the reward size shifted to 1 on day 12, pre-housing × post-housing affected the latency ($P<0.001$). Post hoc analysis showed that again, BB pigs were slower than other pigs (BB: 88.2±2.7; BE 62.3±2.3; EB: 57.3±2.3; EE: 70.4±2.6, $P<0.001$), and EB pigs were faster than EE pigs ($P<0.05$). Pigs ran slower after than before the reward shift ($P<0.001$). Nevertheless, pigs in the 6-1 group ran slower than pigs in the 1-1 group after the reward shift (6-1: 73.9±2.0; 1-1: 66.4±1.8, $P<0.05$), suggesting that pigs are sensitive to a loss in reward size. This was, however, irrespective of housing given the lack of interactions with reward size. We conclude that housing affected the latency to run down a runway for a reward in pigs, which can indicate a lower motivation in the BB pigs, an effect that was absent in the B pigs that switched to enriched housing (BE pigs). We found, however, no evidence that housing or a switch in housing conditions affected the sensitivity to reward loss.

**Effects of dietary L-glutamine as an alternative for antibiotics on the behavior and welfare of weaned pigs after transport**

*Severine P. Parois[1,2], Alan W. Duttlinger[3], Brian T. Richert[3], Jay S. Johnson[1] and Jeremy N. Marchant-Forde[1]*
[1]*USDA-ARS, Livestock Behavior Research Unit, Creighton Hall, 270 S Russell Street, 47907 West Lafayette, USA,* [2]*PEGASE, Agrocampus Ouest, INRA, 16 Le Clos, 35590 Saint-Gilles, France,* [3]*Purdue University, Animal Sciences, Creighton Hall, 270 S Russell Street, 47907 West Lafayette, USA; jay.johnson@ars.usda.gov*

A decrease in antibiotic use in livestock production is a societal demand. Alternative feed supplements have shown promising effects in terms of performance, but their impacts on welfare have had little evaluation. The objectives of the study were to evaluate the effect of no antibiotic, L-glutamine or antibiotic supplementation after weaning and a transport stress on short-term and long-term welfare indicators and behaviors to determine if L-glutamine could be a viable antibiotic alternative for piglets post-weaning. A total of 240 piglets, from 32 distinct litters, were weaned at 18±4.2 days (means ± SD; weight 5.4±1.4 kg) and immediately herded up ramps into a trailer for a 12 h transport. After transport, they were divided, per group of 8, between 3 diets fed for a 2 week period: A – an antibiotic diet including a common commercial prophylactic antibiotic (n=80; chlortetracycline(0.40 g/kg) + Tiamulin(0.035 g/kg)); NA – a control diet without any prophylactic antibiotic or feed supplement (n=80); GLU – a diet including L-glutamine (n=80; 0.20% L-glutamine as-fed). After the 2-week period, all piglets were fed the same control diet. Tear staining, as an indicator of stress, was measured on days 0, 2, 7, 15, 21, 28, 34, 47, 84, 110, 147 post-weaning from photographs. Skin lesions, as an indicator of aggression, were counted 2 days before weaning and on days 2 and 36 post-weaning. Novel object tests (NOT) were done in groups in the pigs' home pen on days 16, 46, 85, 111 post-weaning. The treatment effects were evaluated using repeated measures ANOVA, with treatment groups, periods and their interaction as fixed effects, and the pen as a random effect. The statistical unit was the pig. The NA pigs tended to have larger tear stains than A pigs on days 84, 110 ($P<0.1$), and larger tear stains than GLU pigs on day 84 and 110 ($P<0.05$). Effects on skin lesions were only found on day 2 for the establishment of the hierarchy, with NA pigs (25.3±12.3) having more lesions than A (21.1±12.3) and GLU (18.8±12.0) pigs ($P<0.01$). During the NOT on day 16, A pigs tended to avoid (1.6±2.6 times) the object more than NA pigs (0.8±1.4) ($P<0.1$). On day 85, NA pigs spent less time (190±68.5 s) exploring the object than the two other groups (215±42.7 and 216±41.6 s, both $P<0.01$), and on day 111, NA pigs took longer (11±21.4 s) to interact with the object than GLU (4±5.5 s, $P<0.01$) and A pigs (5±5.7 s, $P<0.05$). The results demonstrate that short-term feeding strategy can have both short- and long-term effects on behavior and welfare. NA pigs appeared less interested in novel objects and more sensitive to environment and management than pigs on the other two treatment groups. Supplementation with L-glutamine appears to confer similar benefits to, and thus could be a viable alternative to dietary antibiotics.

## Examining the relationship between feeding behaviour and sham chewing in nulliparous group-housed gestating sows

*Rutu Acharya, Lauren Hemsworth and Paul Hemsworth*
*University of Melbourne, Animal Welfare Science Centre, 21 Bedford street, North Melbourne,*
*VIC 3051, Australia; rutu.acharya@unimelb.edu.au*

In most modern intensive pig production systems, gestating sows are fed a restricted, highly concentrated diet, delivered over a number of feed drops throughout the day. This type of feed restriction causes reduction in feed intake (approximately 60%) likely resulting in increased hunger, and the reduction in feeding/foraging behaviours which may result in increased frustration and feeding motivation. While the causation of stereotypies such as sham chewing in sows is not fully investigated, there is evidence that stereotypies in sows are strongly related to feeding motivational processes. The present study aimed to examine the relationship between feeding behaviour and sham chewing in nulliparous group-housed gestating sows. Archive video footage of 170 Large White × Landrace group-housed sows in their first gestation was utilised. Sows were floor fed a standard commercial gestation pelleted diet with a daily allowance of 2.5 kg/sow/day over four feed drops per day and water was ad libitum. Sham chewing was defined as repetitive jaw movement without contact with any substrate or feed, and as such the 15-min period post-feed drop was not sampled. Sham chewing of individual sows was recorded using instantaneous point sampling at 30-s intervals from 07:00 to 16:00 h, at days 8 and 52 of gestation. At each observation point a sow was recorded as 'visible' if the sow's snout and jaw were clearly visible or 'not visible' if not viewed clearly. If visible, it was also recorded whether the sow performed sham chewing. Sows were then classified as showing sham chewing at <5% of visible observations, 5-10% of visible observations, and >10% of visible observations, on days 8 and 52 of gestation, and overall. The feeding behaviour of individual sows was recorded using instantaneous point sampling at 30 s intervals over two, 5 min time blocks commencing 30 s after each feed drop, at days 2, 9 and 51 of gestation. A sow was recorded as feeding if she was rooting the ground/feed, or if she was observed to be chewing food. Pearson product moment correlations were used to examine the relationship between feeding behaviour and sham chewing. Data on feeding behaviour was also analysed using a GLMM that included replicate, group/pen (nested within replicate), sham chewing classification, and their interaction as fixed effects. Sham chewing was recorded in 97% of the 170 sows observed on day 8, and 91% of the 150 sows observed on day 52 of gestation. Sows on average were observed performing sham chewing at 10% (SD=0.09) of the visible observation points. A significant negative relationship was found between the proportion of sham chewing on day 52 of gestation and total feeding behaviour on day 9 of gestation (r=-0.24, $P<0.001$). Sows that performed more feeding behaviour on day 9 of gestation performed less sham chewing on day 52 of ($P<0.01$). The present study provides early evidence on relationships between sham chewing and feeding behaviour in group-housed sows, however, as the aetiology of sham chewing is largely unknown further research particularly in older parity sows, or an experimental study with controlled feeding is warranted.

**Use of a commercially available tri-axial accelerometer for cows to monitor sow behavior prior to farrowing**

*Thomas D. Parsons and Mary Jane Drake*
*University of Pennsylvania School of Veterinary Medicine, Swine Teaching and Research Center,*
*382 West Street Road, 19348 Kennett Square, Pennsylvania, USA; thd@vet.upenn.edu*

Video capture and analysis are commonly used to study sow behavior, but it is time-consuming and can be subjective. The specific aim of this study is to validate a commercially available accelerometer for cows in sows in order to study sow behavior in farrowing pens in the peri-parturient period. Yorkshire-Landrace sows (DNA line 241) parity 2-5 were included in the study. IceTag accelerometers were attached to the distal limb of sows. The accelerometers logged movement in three axes 16 times per second and a proprietary algorithm automatically calculated motion index, steps taken, and time spent lying or standing. For validation studies, 3 group-housed gestating sows and 3 individually housed sow immediately pre-partum were observed for a total of 328 min using either a handheld camcorder or fixed overhead cameras. Recorded video was compared to posture and step data output from the accelerometers in order to determine their accuracy and sensitivity. Following validation, the IceTags were used to monitor the behavior of 11 sows across 3 different farrowing groups in the peri-parturient period from the time sows entered the farrowing room until 48-72 hours post-farrowing. Straw was provided to 6 sows ~36 hours prior to farrowing. Comparison of video and accelerometer data revealed that the latter were able to accurately capture sow locomotion in both open housing situations (gestation) as well as in more confined farrowing pens. A regression line fitted to the scatter plot of step counts using the two methods had a high degree of correlation ($F(1,318)=25,840$, $P<0.001$, m=0.965, $R^2=0.983$). During validation, 1,176 steps were recorded by the accelerometer and 1,160 steps observed in the video (>98% overall consistency) and the absolute value of the difference between the two approaches summed to 114 steps (<10% absolute error). As the accelerometers are able to detect posture (lying vs standing), we affixed accelerometers to both the forelimb and hindlimb of the same sow in order to assess additional postures (sitting, kneeling) by comparing the reported posture of the two limbs. This also provided more resolution in analyzing the frequency of postural changes. Using this configuration, an increase in the activity levels of sows was observed beginning approximately 24 hours before farrowing. Activity levels peaked at 4-6 hours before farrowing as the frequency of postural changes increased to 15-25 changes/hr, compared with 0-5 changes/hr at baseline (>48 hr before farrowing). This was related to a ~4-fold increase in mean posture changes ($t(11)=1.72$, $P<0.001$) in the 12 hours prior to farrowing compared to baseline (>48 hr). A linear mixed model revealed that time prior to farrowing, but not access to straw, affected the number of postural changes per hour ($\chi^2(1)=119.6$, $P<2.2e^{-16}$). These activity increases are consistent with behavior changes observed in sows during the nest building phase of labor. The IceTag technology is used for heat detection and health monitoring in cows. Results from this study indicate that these accelerometers can also be reliably used for monitoring sow locomotion and posture.

## Effects of social status on sow behaviour, enrichment use, and cortisol concentrations in group housed gestating sows

Fiona C. Rioja-Lang[1], Victoria Kyeiwaa[2,3], Laurie M. Connor[4], Yolande M. Seddon[5] and Jennifer A. Brown[2,3]
[1]The University of Edinburgh, JMICAWE, Easter Bush, Midlothian, EH25 9RG, United Kingdom, [2]The University of Saskatchewan, Dept. of Animal & Poultry Science, 51 Campus Drive, Saskatoon, Saskatchewan, S7N 5A8, Canada, [3]Prairie Swine Centre, Ethology Group, Box 21057, 2105 – 8th Street East, Saskatoon, Saskatchewan, S7H 5N9, Canada, [4]The University of Manitoba, Dept. of Animal Science, 201-12 Dafoe Road, Winnipeg, Manitoba, R3T 2N2, Canada, [5]The University of Saskatchewan, WCVM, 52 Campus Dr., Saskatoon, Saskatchewan, S7N 5B4, Canada; fiona.lang@ed.ac.uk

The current Canadian Code of Practice for Care and Handling of Pigs requires that all new or renovated housing for gestating sows use group housing. Providing environmental enrichment can reduce aggression and improve the welfare of sows in groups. However, if enrichment is considered a valuable resource, then dominant sows may obtain greater access than subordinates. The objective of this study was to examine the effects of social status on sow behaviour, enrichment use, aggression, and cortisol concentrations in a group system. Eight groups of 27 ($\pm$1) sows were observed, with additional data from six focal sows (3 dominant [Dom] and 3 subordinates [Sub]) per group, identified using a feed competition test. Each group received four treatments in random order, for 14 d each. Treatments were: (1) constant provision of wood enrichment; (2) rotation of three objects (rope, straw, and wood); (3) rotation of three objects with an associative stimulus; and (4) no objects (control). Photographs were taken every 10 min on days 1, 8, 10 and 12, to determine enrichment use and postures of sows. Stereotypic behaviours were observed in Dom and Sub sows using live observations by a single observer every 5 min for 60 min in the morning and afternoon. Initial and final body weights of sows were recorded, and aggression was evaluated using skin lesion scores on a scale of 0 (no injury) to 3 (severe injury). Saliva samples were collected from focal sows at 5, 9 and 14 weeks of gestation. Saliva was collected by allowing sows to chew on cotton balls wrapped on metal rods until thoroughly moistened. Data were analyzed using Proc Mixed and Proc Glimmix in SAS 9.4. Dom sows were of higher average parity (3.11$\pm$0.09; 2.89$\pm$0.08 respectively, mean $\pm$ SD) and significantly heavier (256.5; 239.3 kg; SEM$\pm$5.35. $P<0.01$ respectively) than Sub sows. A greater number of Dom sows were observed active and standing for more time ($P<0.05$) and displayed a greater amount of bar biting than Sub sows ($P<0.05$). On average, sows were observed in contact and within one meter of enrichment in 74% of observations on day 10 compared to 50%, 46 and 57% of observations on days 1, 8 and 12 respectively. However, enrichment use did not differ due to social status ($P>0.05$). Social status did not influence aggression or cortisol concentration (Dom=0.18; Sub=0.19 µg/dl; $P>0.05$). However, cortisol concentrations from weeks 5 and 9 (early/mid gestation) were significantly lower than those taken on week 14 (late gestation) (0.14; 0.14; 0.33 µg/dl; $P<0.05$). In conclusion, there were similar levels of enrichment use, and no difference in lesions or physiological response among Dom and Sub sows, rather, stage of gestation influenced cortisol concentration.

**Effects of enrichment type (object and fibre) and number on group housed sows with dominant and subordinate social status**

*Cyril Roy[1], Yolande Seddon[2], Jennifer Brown[1] and Laurie Connor[3]*
[1]*Prairie Swine Centre, 2105- 8th Street East, P.O. Box 21057, S7H 5N9, Saskatoon, Saskatchewan, Canada,* [2]*Western College of Veterinary Medicine, 52 Campus Drive, Saskatoon, SK, S7N 5B4, Canada,* [3]*University of Manitoba, Animal Science, 12 Dafoe Rd, Winnipeg, MB, R3T 2N2, Canada; cyril.roy@usask.ca*

The Canadian Code of Practice for the Care and Handling of Pigs requires that all pigs be provided with multiple forms of enrichment. However, what types of enrichment will maximize utilization and welfare benefits in sows are not well understood. This study examined five enrichment treatments to determine their effects on enrichment use and aggression in dominant and subordinate sows. A total of 140 sows (5 replicates in static groups of 28) were housed in gestation pens with free-access stall feeding and space allowance of 2.2 $m^2$ per sow. The enrichments provided in the lounging area were; 1 or 3 hanging wood objects (W1, W3); 1 or 3 hoppers dispensing hay (H1, H3) and a control treatment (C; no enrichment). Each replicate received three of the five treatments in random order, with each treatment lasting for 19 days. Groups were formed at 5 weeks gestation and sows entered the trial six days after mixing. Video camera footage of 3 sessions of 1 hour (morning, afternoon and evening) was obtained on days 1, 5, 12 and 19 of each treatment. Scan samples taken at 10-minute intervals were used to determine enrichment use, pen location and postures of sows at group level. Based on a feed competition test (8 days after mixing), six focal sows (three dominant [D] and three subordinate [S]) were selected per group to study effects of social status on enrichment use, aggression (skin lesions: scale of 0-3 in 8 body regions) and health (body weight, back fat and body condition). PROC GLIMMIX and PROC MIXED were used to find associations between outcome variables and different treatments, with replicate as random effect. The four enrichment options provided had no effect on the number of sows contacting enrichment. However, more sows were in proximity to enrichment (% of sows <1 m from enrichment) when hay was provided, (avg H1, H3: 2.75±0.20%) than when wood enrichments were given (avg W1, W3: 0.56±0.20%; LSmeans ± SEM, $P<0.001$). Among focal animals, fewer D sows were found in stalls than S sows (82.15±0.97 vs 90.29±0.97%; $P=0.001$). Conversely, D sows laid more in the enrichment area than S sows (9.47±1.2 vs 3.80±1.2%; $P=0.001$). There was no effect of social status on skin injuries measured on days 1, 5, 12 or 19. On days 5 and 12, sows on H1 and H3 treatments had a significantly higher number of injuries compared to wood and control, indicating more competition (Day 5; avg H1, H3=14.81±0.41 vs avg W1,W3=4.23±0.41 and C=4.66±0.41; $P=0.01$). Sows in the W1 treatment had the fewest skin injuries (avg 2.46 injuries for the 4 days measured). Enrichment treatment and social status did not show any significant association with health variables. In conclusion, sows spent more time near enrichments when fibre was provided probably leading to more injuries. D sows spent more time near enrichment than S sows. Provision of a single wood enrichment resulted in fewer injuries.

**A preliminary examination of sham chewing across first and second gestation in group-housed sows**

*Maxine Rice, Rutu Acharya, Lida Abrie, Paul Hemsworth and Lauren Hemsworth*
*Animal Welfare Science Centre, The University of Melbourne, North Melbourne VIC, 3051, Australia; mrice@unimelb.edu.au*

Stereotypies are repetitive behaviours induced by frustration, repeated attempts to cope, and/or CNS dysfunction. The barren environments often associated with modern intensive pig production systems have been implicated in the development of stereotypic behaviours such as sham chewing. Sham chewing is the most common and frequently observed stereotypy in group-housed sows. The origins, characteristics and welfare implications of stereotypic behaviour in group-housed sows have received little examination and remain largely unknown. This preliminary study is part of a larger project investigating the relationships between sham chewing and the welfare and productivity of group-housed gestating sows. It aimed to examine the performance of sham chewing across first and second gestation in a small sample of group-housed sows. Archive video footage of 19 Large White × Landrace group-housed sows in their first and second gestation was used. Sows were twice artificially inseminated and within seven days of insemination randomly mixed into groups of ten (floor space of 1.8 $m^2$/gilt). Sows were floor fed a standard commercial gestation pelleted diet with a daily allowance of 2.5 kg/sow/day over four feed drops per day. Water was ad libitum. From video recordings, observations of sham chewing of individual sows were conducted using instantaneous point sampling at 30-s intervals from 07:00 to 16:00 h at day 8 of gestation. Sham chewing was defined as repetitive jaw movement without contact with any substrate or feed, and as such the 15-min period post-feed drop was not sampled. At each observation point, a sow was recorded as 'visible' if the sow's snout and jaw were clearly visible or 'not visible' if the sow's snout and jaw were not able to be viewed clearly. If visible, a sow was also recorded as either performing sham chewing or not performing the behaviour. Productivity outcomes for each sow were measured across both gestations; total live born, total still born, and total piglets weaned. Paired sample t-tests were used to examine differences in sham chewing between the two gestations. Sham chewing was recorded in 100% of sows observed on day 8 of gestation 1 and gestation 2. The mean proportion of visible observation points where sows were performing sham chewing behaviour on day 8 of gestation 1 and gestation 2 were 0.12 (SD=0.09) and 0.07 (SD=0.08), respectively. There was no difference in the proportion of visible observation points in which sham chewing was shown by sows on days 8 in their first and second gestation ($t_{18}$=1.67, P=0.11). There were no differences in productivity between the two gestations; total live born ($t_{11}$=1.04, P=0.31), total still born ($t_{11}$=-1.0, P=0.34), and total piglets weaned ($t_{11}$=-1.42, P=0.18). The present study found no significant difference in a sow's performance of sham chewing or her productivity outcomes between first and second gestation. However, given this preliminary study involves a small number of animals and the limited literature suggests that both the frequency and duration of stereotypic behaviour increases with parity, further research examining sham chewing across gestation in multiparity group-housed sows is warranted.

## Nutritional value of pasture and dry-forage, and foraging behaviour in outdoor gestating sows

*Lydiane Aubé[1,2], Frédéric Guay[1], Renée Bergeron[3], Gilles Bélanger[4], Gaëtan Tremblay[4], Sandra Edwards[5], Jonathan Guy[5], Jérôme Théau[6] and Nicolas Devillers[2]*
[1]*Université Laval, 2425, rue de l'Agriculture, Québec, Canada,* [2]*Agriculture and Agri-Food Canada, 2000 College Street, Sherbrooke, Canada,* [3]*University of Guelph, 50 Stone Road East, Guelph, Canada,* [4]*Agriculture and Agri-Food Canada, 2560 Hochelaga Blvd, Québec, Canada,* [5]*Newcastle University, King's Road, Newcastle, United Kingdom,* [6]*Université Sherbrooke, 2500 Blvd de l'Université, Sherbrooke, Canada; lydiane.aube@canada.ca*

The aim of this study was to determine if forage intake could contribute up to 60% of the nutrient needs of gestating sows kept outdoors. A total of 45 sows were distributed among three treatments (5 replicates of 3 sows/treatment) from 5 weeks of gestation until farrowing. Treatments differed in the daily level of concentrate feed in terms of percentage of metabolic energy (ME) and the type of forage offered during gestation as follows: T1, 90% of ME needs and access to a pasture plot (25×50 m, legume-grass mixtures); T2, 40% of ME needs and access to a pasture plot, and T3, 40% of ME needs and access to a bare paddock and hay ad libitum. From farrowing to weaning (5 weeks), sows were fed concentrate ad libitum. Body weight and back fat depth were measured during gestation and lactation. Postures of sows were assessed over three 48 h periods at the beginning, middle, and end of gestation. Hay intake (T3) was evaluated by weighing hay refusal every two weeks. Fresh forage intake (T1 and T2) was calculated as the difference of biomass before and after grazing pasture plots, corrected for forage growth. Biomass was estimated using the green normalized difference vegetation index (GNDVI) calculated from aerial images. Data were analysed with repeated measures according to a generalized complete block design. Post-hoc comparisons (Tukey's adjustments of Student's t-test) between treatments were done for each period when interaction treatment×period was significant. At the end of gestation, T1 sows were heavier and had greater back fat thickness than T2 and T3 sows (290, 259, and 250 kg, $P<0.001$; 20.1, 16.6, and 16.0 mm, $P=0.002$; respectively). During lactation, T1 sows maintained a higher body weight ($P=0.004$) and tended to have greater back fat thickness ($P=0.09$) than T3 sows, but differed from T2 sows only on d1 of lactation for body weight ($P<0.001$) and back fat ($P=0.012$). At weaning, T3 sows tended to be lighter than T2 sows (226 vs 241 kg, $P=0.08$). Fresh forage intake tended ($P=0.09$) to be higher for T2 than T1 sows (5.2 vs 3.9 kg DM/sow/day). Hay intake in T3 sows increased linearly throughout gestation (from 150 to 877 g DM/sow/day, $P<0.001$). The T1 sows spent less time standing than T2 and T3 sows (23, 31 and 27±1.8%, respectively, $P=0.005$) in mid-gestation. At the end of gestation, T1 sows tended to spend less time standing than T2 sows (24 vs 30±1.8%, $P=0.06$). In conclusion, sows fed concentrates to meet 40% of ME needs, did not consume enough fresh forage or hay to maintain the same body condition as sows fed concentrates to meet 90% of ME needs. Despite their inability to fully compensate concentrate reduction, T2 sows seem to have had a higher level of activity than T1, probably due to greater foraging behaviour, and were able to maintain acceptable body condition.

**Testing the feasibility of using a conveyor belt to load weanling and nursery pigs for transportation**

*Donald Lay, Avi Sapkota and Stacey Enneking*
*Agricultural Research Service-United States Department of Agriculture, Livestock Behavior Research Unit, 270 S. Russell Street, Creighton Bldg, Room 3010, West Lafayette, IN 47907, USA; don.lay@ars.usda.gov*

Transportation is known to be a multi-faceted stressor, with the process of loading being one of the most significant factors impacting the stress to which pigs are exposed. This project was designed to determine if using a conveyor to load pigs into the top deck of a simulated straight deck trailer could lower the stress to which pigs and handlers are exposed. Pigs were assigned to either a Control group that were herded up a stationary conveyor ramp into a top deck trailer (2.5 m above the ground); or Conveyor group which were herded onto a mobile conveyor into a top deck trailer. The conveyor was 7.6 m long, 0.9 m wide and rose to 2.5 m high at a 16 ° slope, and moved 11.3 m/min. Two age groups were tested, Weaned pigs (21 d of age) which were moved in groups of 20 (n=14 groups/treatment); and Nursery pigs (48 d of age) which were moved in groups of 10 (n=15 groups/treatment). Group sizes were based on recommendations from the National Pork Board (U.S.). Behavior was recorded during loading, including slips and falls, vocalizations, assists, and time to load. During loading, the heart rate of 2 sentinel pigs/group and the handler, and after loading, the body temperature of the handler were recorded. Pigs were held in the simulated trailer for 30 min while heart rate was recorded. After which, they were unloaded and held in a holding pen for an additional 30 min while heart rate was recorded. Data were analyzed using a mixed model analysis of variance. There were no treatment differences for slips or falls ($P<0.90$). Vocalizations were too few to analyze. Both Weaned ($2.8\pm0.7$) and Nursery ($1.6\pm0.5$) Conveyor pigs needed to be assisted onto the conveyor more than Weaned ($1.2\pm0.4$) and Nursery ($0.3\pm0.1$) Control pigs ($P<0.06$). There was no difference in total loading time between the treatments for any age group ($P<0.15$), with Weaned and Nursery pigs loading in 50 to 45 s as a group, respectively. There were no treatment differences for heart rate variability measures ($P>0.10$). However, loading increased heart rate of Nursery pigs ($204.9\pm5.7$ bpm, $P<0.005$), but not Weaned pigs ($172.1\pm9.0$ bpm). Nursery pigs had a greater ratio of low frequency to high-frequency power ratio of total autonomic activity ($<0.09$ to $>0.09$ Hz) a measure of sympathetic dominance, during loading ($P<0.02$) compared to other phases of the procedure in both Control and Conveyor groups. Heart rate ($93.9\pm1.9$ bpm) and body temperature ($31.1\pm0.3$ °C, eye temperature) of the handler were not affected by treatment ($P<0.26$). Based on behavior and physiology, the pigs had similar experiences in both treatments. This study shows that it is feasible to use a conveyor to load pigs, but it may not be advantageous.

## Goat kids display minimal aversion to relatively high concentrations of carbon dioxide gas

*Marcus McGee[1,2], Rebecca L. Parsons[2], Mhairi A. Sutherland[3], Alejandro Hurtado-Terminel[2], Paul J. Plummer[2] and Suzanne T. Millman[2]*
*[1]Northwest Missouri State University, Agricultural Sciences, Maryville, MO 64468, USA, [2]Iowa State University, Veterinary Diagnostic & Production Animal Medicine, 1809 S Riverside Dr., Des Moines, IA 50011, USA, [3]AgResearch, Ruakura Research Centre, Hamilton, 3240, New Zealand; smillman@iastate.edu*

$CO_2$ euthanasia may only be used where aversion and distress can be minimized. Our objectives were to assess aversion to $CO_2$ by goat kids using approach-avoidance and conditioned place avoidance paradigms. We hypothesized kids would avoid $CO_2$ concentrations >30%, and display conditioned place avoidance if loss of consciousness (LOC) occurred. A preference-testing device consisted of 2 chambers separated by a sliding door and exhaust sink. Twelve kids were trained for 5 d to enter the treatment chamber (TC) once daily to access the milk ration, with chambers maintained at 1% $CO_2$. After entry, kids had 5 min to move freely between chambers. During testing, TC was maintained at 1 of 4 $CO_2$ concentrations: 30, 50, 70 or 90%. Tests concluded when LOC occurred or after 5 min, and kids systematically received each $CO_2$ treatment over 4 rounds. Each round consisted of 1% $CO_2$ on D-1, assigned $CO_2$ treatment on D0 and 1% $CO_2$ on D+1. Behavior outcomes were collected from video and live observations. PROC PHREG of SAS was used to assess differences, using a model including day, $CO_2$ treatment, and interactions, with round as covariate; PROC GLIMMIX was used to calculate lsmeans. All kids entered TC on all D0 days, except 6 kids refused at 90% $CO_2$. Kids that entered TC on D0 consumed milk prior to and during ataxia; all displayed LOC. On D0, latency to enter TC was greater at 90% $CO_2$ (n=6; mean ± SE=29.3±5.0 s) than other $CO_2$ treatments ($P<0.01$; 6.5±2.3 s (30%), 8.6±2.7 s (50%), 9.2±2.8 s (70%). For 90% $CO_2$, latency to enter was greater on D+1 ($P<0.01$; 10.3±2.8 s) than D-1 (4.4±1.8 s), suggesting conditioned place aversion. There was a negative relationship with round ($P<0.01$), such that kids entered TC faster through rounds of testing. In conclusion, kids did not display aversion to $CO_2$ at 70% concentration or less.

## A two-step process of nitrous oxide before carbon dioxide for humanely euthanizing piglets: on-farm trials

*Rebecca Smith[1], Jean-Loup Rault[2], Richard Gates[3] and Donald Lay[4]*
[1]*Purdue University, Department of Animal Sciences, 270 S. Russel Street, West Lafayette, IN 47907, USA,* [2]*Institute of Animal Husbandry and Animal Welfare, University of Veterinary Medicine, 1210, Vienna, Austria,* [3]*University of Illinois, Dept. of Agriculture and Biological Engineering, 1304 W. Pennsylvania Avenue, Urbana, IL 61801, USA,* [4]*USDA-ARS, Livestock Behavior Research Unit, 270 S. Russel Street, West Lafayette, IN 47907, USA; smit1934@purdue.edu*

Current methods of euthanizing neonatal piglets on-farm are raising animal welfare concerns. Nitrous oxide ($N_2O$), also known as laughing gas, is widely used in human medicine as an anesthetic, has been found to work as an anesthetic in piglets, and using it for anesthesia prior to carbon dioxide ($CO_2$) may be a more humane method than using $CO_2$ alone. Our experiment was the first to test the use of a two-step euthanasia method, using $N_2O$ for six minutes and then $CO_2$ until death on ill and/or compromised neonate piglets. A commercial euthanasia chamber was modified to deliver these two euthanasia treatments: the two-step delivery of $N_2O$ followed by $CO_2$ to death or only $CO_2$ from start to death. Gases were delivered at a 25% replacement rate per minute. In Experiment 1, eighteen piglets were individually euthanized (n=9), and in Experiment 2 eighteen groups of 4-6 piglets were euthanized (n=9). Behavior data (postures and activities) were collected during the processes on 0- to 7-day-old piglets using a camera and digital sound recorder. Data were analyzed in SAS with the piglet or group of piglets as the experimental unit. Normal data were analyzed as a mixed model analysis of variance and non-normal data were analyzed using the Wilcoxon-Mann-Whitney test statistic. For both experiments, the sequence of behaviors was different. In the two-step $N_2O/CO_2$ treatment, piglets lost posture (Exp. 1: 284.36±43.29 seconds, Exp. 2: 191.50±11.61 seconds), an indicator of the onset of losing consciousness, before going into distressful behaviors of heavy breathing (Exp. 1: 356.79±31.29 seconds, Exp. 2: 364.46±5.35 seconds) and open-mouth breathing (Exp. 1: 439.87±9.99 seconds, Exp. 2: 386.22±4.44); whereas piglets in the $CO_2$-only treatment did not lose posture (Exp. 1: 125.16±6.48 seconds, Exp. 2: 71.80±5.13) until after exhibiting these distressful behaviors (heavy breathing = Exp. 1: 61.61±10.93, Exp. 2: 30.94±2.15; open-mouth breathing – Exp. 1: 86.71±9.82, Exp. 2: 39.94±2.58; $P\leq0.004$). Heavy breathing and open-mouth breathing are signs of breathlessness and are associated with unpleasantness and compromised welfare. However, piglets in the $N_2O$ treatment took longer to lose posture compared to the $CO_2$ treatment ($P<0.001$). The $N_2O$ treatment displayed more squeals/minute ($N_2O$: 4.90±1.41, $CO_2$: 0.48±0.25, $P=0.004$), escape attempts per pig ($N_2O$: 0.79±0.32, $CO_2$: 0.17±0.12, $P=0.021$), and righting responses per pig ($N_2O$: 1.41±0.39, $CO_2$: 0.36±0.06, $P=0.084$). Squeals, escape attempts, and righting responses are considered behavioral signs of stress and aversion. In these regards, it cannot be concluded that euthanizing piglets for six minutes with $N_2O$ and then $CO_2$ is more humane than just euthanizing with $CO_2$.

## Aversion to carbon dioxide gas in turkeys using approach-avoidance and conditioned place avoidance paradigms

*Rathnayaka M.A.S. Bandara[1], Stephanie Torrey[1], Rebecca L. Parsons[2], Tina M. Widowski[1] and Suzanne T. Millman[2]*
[1]*University of Guelph, Animal Biosciences, 50 Stone Road East Guelph, Ontario, N1G 2W1, Canada, [2]Iowa State University, Department of Veterinary Diagnostic & Production Animal Medicine, 2203 Lloyd Veterinary Medical Center, Ames, IA 50011, USA; rbandara@uoguelph.ca*

Carbon dioxide is widely used for poultry euthanasia. To claim $CO_2$ as a human euthanasia agent, it is essential to identify $CO_2$ concentrations which are not aversive to the species and cause rapid insensibility. The objective of this study was to examine the aversiveness of $CO_2$ to turkeys (2-6 h food deprived) using approach-avoidance and conditioned place avoidance paradigms. A preference test was designed using two identical chambers separated by a sliding door and a curtain. The control chamber (CC) maintained ambient air conditions (1% $CO_2$); the treatment chamber (TC) maintained predetermined $CO_2$ concentrations. Eleven turkeys (1.26±0.18 kg; 4 wks old) were individually trained for 5 consecutive days to enter the TC by pushing through a curtain to access rewards (feed and enrichment) with ambient air in both chambers. After 5 minutes in the CC, the sliding door was opened to provide access to the TC. Birds received 5 minutes access to the TC, after which they were removed and returned to the home pen. All birds learned to enter TC and were advanced to testing phase after two consecutive days of entering to TC. During the testing phase, the same procedures were used with the TC maintained at one of four $CO_2$ levels: 25, 35, 50 or 70%. Tests were concluded after 5 minutes or when loss of neck tone (LONT) occurred during gas exposure and birds were removed from the chamber for recovery. Turkeys experienced each of the $CO_2$ treatments (in random order) on gas day (G), preceded by one baseline day (B, 1% $CO_2$) and followed by one wash-out day (W, 1% $CO_2$).The same rewards were provided in the TC for all B,G, and W. Four birds never entered TC after the first gas experience. According to Fisher's exact test, there is a significant difference ($P<0.0001$) in the proportion of birds that entered TC on G for the different levels of $CO_2$. All 8 turkeys tested entered TC at 25% $CO_2$. Only three birds out of 8 entered TC for the 35% level and only one bird out of seven entered TC for the 50% $CO_2$. Eight birds were tested for 70%, and none entered TC. Pairwise comparisons (2×2 Fisher's exact test with Bonferroni correction) showed that the proportion of birds that entered TC on G was significantly different between 25% and all other $CO_2$ levels (6 pairs = 25 vs 35%: $P=0.026$, 25 vs 50%: $P<0.001$, 25 vs 70%: $P<0.001$, 35 vs 50%: $P=0.559$, 35 vs 70%: $P=0.07$, and 50 vs 70%: $P=0.46$). LONT followed Loss of Posture (LOP) in all birds that entered TC at all $CO_2$ levels (average latency ± SD; 25%: LOP=18±7 s, LONT=49±14 s; 35%: LOP=15±2 s, LONT=38±7 s; 50%; LOP=8 s, LONT=15 s). Conditioned place avoidance was observed in 2 birds exposed to 25% $CO_2$, and not for other concentrations. None of the turkeys tested actively avoided 25% $CO_2$.These results suggest that turkeys do not find 25% $CO_2$ aversive and that exposure at this level causes LONT, indicative of loss of consciousness within less than one minute.

**Variability in the emotional responses of rats to carbon dioxide**
*Lucía Amendola, Anna Ratuski and Daniel M. Weary*
*University of British Columbia, Animal Welfare Program, 2357 Main Mall, Vancouver, V6T1Z4*
*British Columbia, Canada; luciaame@mail.ubc.ca*

Carbon dioxide, commonly used to kill laboratory rats, is known to be aversive even at concentrations far lower than that required to render animals unconscious. However, aversion is variable among individuals. Humans also vary in their responses to $CO_2$; approximately 50% of healthy participants experience anxiety following a double inhalation between 9 and 35% $CO_2$, but panic disorder patients experience anxiety with a single inhalation. These emotions experienced are repeatable and attenuated by anxiolytics. The aims of this study were to assess repeatability and consistency of individual rat thresholds of aversion, and to evaluate the effect of midazolam (an anxiolytic) on aversion. Aversion was measured by direct observation, as latency to avoid $CO_2$ gradual fill at the cost of losing a sweet food reward (approach-avoidance test). Female Sprague Dawley rats were tested six times at each of three ages: 3 (n=9), 9 (n=9), and 16 months (n=6). During the last three tests at age 16 months, rats were pre-treated with midazolam (0.32 mg/kg). The individual rat was used as the experimental unit. Repeatability (R) was estimated at each age. The mean latency to avoid $CO_2$ per rat within each age was used to assess consistency between consecutive ages using Pearson correlation (r). The effect of anxiolytics on $CO_2$ aversion was assessed using Linear Mixed Models with repetition and treatment (pre and during anxiolytics) as fixed factors, and rat as random intercept. Aversion was repeatable at 3 (R=0.39), 9 (R=0.63) and 16 months (R=0.47), and was consistent from 3 to 9 months (r=0.65) and from 9 to 16 months (r=0.71). Latency increased by 61% when rats were treated with midazolam ($F_{1,26}=22.9$, $P<0.001$). These suggest that rats likely experience anxiety during $CO_2$ exposure and that individuals vary in this response.

## Vets, owners, and older cats: exploring animal behavioural change in euthanasia decision-making

*Katherine E. Littlewood[1], Ngaio J. Beausoleil[1], Kevin J. Stafford[1] and Christine Stephens[2]*
[1]*Massey University, School of Veterinary Science, Private Bag 11-222, Palmerston North 4442, New Zealand,* [2]*Massey University, School of Psychology, Private Bag 11-222, Palmerston North 4442, New Zealand; k.littlewood@massey.ac.nz*

As veterinary care for pet cats has improved, many of these animals are living longer. For various reasons, some owners believe that an extended life is preferable to a humane death. Such concerns are important for older cats and those with chronic disease because these animals tend to deteriorate slowly and may be kept alive past the point at which they are experiencing an 'acceptable' quality of life. To date, surveys and structured interviews have formed the basis of research into this area of animal welfare concern. While these quantitative methods are useful for investigating population-level features of end-of-life decision-making, qualitative approaches, such as in-depth interviews, are valuable for the rich detail they provide. This presentation will discuss the qualitative research methods used and indicate some preliminary findings from thematic analysis of interviews with New Zealand cat owners and their veterinarians. Our study set out to explore the vet's role in end-of-life management of chronically ill and older cats. This involved analysing how the owner made the decision to end their cat's life and the role of the vet in this process. The decision to end the life of a beloved pet, or family member, is difficult and involves many factors. Taking owner's observations of their cat's behaviour into account is one factor that can assist with such decisions. Systematic interviews were conducted with owners that had made the decision to euthanase their older or chronically ill pet cat within the last six months. The veterinarian involved in the euthanasia event was also interviewed separately. There were fourteen owner-vet interview pairings. Each interview was digitally recorded and transcribed. Thematic analysis was performed on interview transcripts. Thematic analysis has revealed that many cat owners identified behavioural changes in their pet that they believed indicated a declining quality of life. Many of these behavioural changes reflected a reduction in play or perceived enjoyment for the cat. Conversely, there were also subtle behavioural changes that were difficult for owners to detect, e.g. reduced grooming resulting in poor coat condition. In these cases, it often took a third party (e.g. friend or vet), who had not seen the cat recently, noticing physical changes to instigate discussion of behavioural changes. Some vets were aware of these behavioural changes in their client's animals, while others were not. However, for many owners, a change in their pet's behaviour was a significant factor in their decision to end its life. Thematic analysis confirmed that behavioural changes are often the first thing an owner notices about their cat's declining quality of life. This qualitative approach has provided a deeper understanding of ways in which behavioural changes have been used to inform end-of-life decisions and emphasises the importance of questions asked by vets that focus owners on changes in an elderly or chronically ill pet's behaviour.

## Comparison of EEG and behavioural changes for on-farm euthanasia of turkeys

*Elein Hernandez[1], Fiona James[2], Stephanie Torrey[3], Tina Widowski[3] and Patricia V. Turner[1]*
[1]*University of Guelph, Pathobiology, 50 Stone Rd E, N1G 2W1 Guelph, ON, Canada,* [2]*University of Guelph, Clinical Studies, 50 Stone Rd E, N1G 2W1 Guelph, ON, Canada,* [3]*University of Guelph, Animal Biosciences, 50 Stone Rd E, N1G 2W1 Guelph, ON, Canada; elein@uoguelph.ca*

There is a critical need to develop appropriate on-farm euthanasia methods for all poultry species. Preferred methods of euthanasia should affect the brain first, resulting in rapid loss of sensibility, followed by cessation of cardiac and respiratory activity. Intravenous (I.V.) barbiturate overdose is considered the gold standard method of euthanasia method for all species, but effects have not been specifically reported for turkeys. Cervical dislocation (CD), either manual or mechanical, is a conditionally acceptable method with several welfare concerns including the potential for prolonged time to of sensibility and death following application. Electroencephalography (EEG) is considered the most objective method for assessing time to brain death for mammalian and avian species, that is associated with electrocerebral inactivity (<2 uV) and observed as an isoelectric (flat) line in the EEG. The aim of this study was to evaluate EEG and behavioural responses of turkeys euthanized with I.V. barbiturate overdose or mechanical or manual CD. EEG and bird behaviour were recorded continuously for a maximum of 6 minutes following euthanasia method application using lifelines ambulatory EEG system. Female turkeys (8 weeks old, mean body weight 2.5 kg) were randomly assigned to treatment groups and euthanized with intravenous pentobarbital sodium (1 ml/4.5 kg)(n=6), manual CD(n=10) or the Koechner euthanasia device (KED, mechanical CD, size medium) (n=12). EEGs were recorded in lightly anesthetized birds (1.5% isoflurane/100% $O_2$ by face mask) with four 25G needle electrodes inserted subcutaneously to the head for baseline activity prior to application of the euthanasia method. Birds euthanized with pentobarbital presented with isoelectric EEG signals approximately 18±46.3 s after I.V. drug administration, with feather erection occurring at 57.9 s. KED and manual CD resulted in prolonged latency to isolectric signal compared to barbiturate group (F 7.35,$P$=0.006). For manual CD, the average time to an isoelectric signal was 238.6±41.4 s with a mean time to onset of clonic and tonic convulsions at 77.6±31.3 and 153.2±29.85 s, respectively. Isoelectric signals occurred in only 6 birds euthanized with the KED, occurring at 268.6±52.4 s, with a mean onset of clonic and tonic convulsions at 135.9±24.3 and 204.9±27.9 s, respectively. The remaining birds in this group were euthanized by another method at 6 min. Cardiac arrest occurred in turkeys within 2-3 min after barbiturate administration whereas cardiac arrest was only noted in one bird in the KED and one in the manual CD euthanasia groups before the 6-minute endpoint. Barbiturate overdose is an effective euthanasia method but with on-farm limitations. Manual CD is a common method of on-farm euthanasia, but personnel training and potentially prolonged brain death are welfare concerns. Use of the KED for turkey poult euthanasia resulted in highly inconsistent onset of brain death and alternative euthanasia methods should be considered for this age group of turkeys.

## Sham chewing and sow welfare and productivity in nulliparous group-housed gestating sows

*Lauren Hemsworth, Rutu Acharya and Paul Hemsworth*
*Animal Welfare Science Centre, The University of Melbourne, North Melbourne VIC, 3051, Australia; lauren.hemsworth@unimelb.edu.au*

Stereotypies are repetitive behaviours induced by frustration, repeated attempts to cope, and/or CNS dysfunction. While it is believed that stereotypies develop due to suboptimal environments and indicate a welfare concern, the actual welfare implications remain poorly understood. Despite the move from stall to group-housing during gestation, stereotypies such as sham chewing are still anecdotally observed in group-housed sows. This study aimed to examine the relationships between sham chewing and the welfare and productivity of nulliparous group-housed gestating sows. Archive video footage of 170 Large White × Landrace sows in their first gestation was utilised. Gilts were randomly mixed into groups of ten (floor space of 1.8 $m^2$/gilt), in partially slatted, non-bedded and non-enriched pens. Sows were floor fed a standard commercial gestation pelleted diet with a daily allowance of 2.5 kg/sow/day over four feed drops/day. From video recordings, observations of sham chewing of individual sows were conducted using instantaneous point sampling at 30-s intervals from 07:00 to 16:00 h, on days 8 and 52 of gestation. Sham chewing was defined as repetitive jaw movement without contact with any substrate or feed. At each observation point, if visible (snout and jaw clearly visible), a sow was also recorded as either performing sham chewing or not performing the behaviour. Based on the proportion of occurrences (i.e. the proportion of visible observation points where sows were performing sham chewing), sows were classified as showing sham chewing at <5% of visible observations (Category 1), 5-10% of visible observations (Category 2), and >10% of visible observations (Category 3). Sow welfare and productivity outcomes were measured; plasma cortisol concentrations on days 2, 9, and 51 of gestation, live weight gain from days 2 to 100 of gestation, litter size and farrowing rate. Cortisol, reproductive success and sham chewing were analysed separately using a GLMM. Sham chewing was recorded in 97% of sows on day 8, and 91% of sows on day 52 of gestation. Sows on average were observed performing sham chewing at 10% of visible observation points. There was no difference in the proportion of visible observation points in which sham chewing was shown by sows on days 8 and 52 ($t_{149}$=1.02, $P$=0.31). There were few significant relationships found between sham chewing and welfare and productivity variables. Sows that performed sham chewing in <5% of visible observations on days 8 and 52 had fewer still-born piglets ($P<0.05$) and lower cortisol concentrations on day 8 ($P<0.05$), than sows performing sham chewing at >5% of visible observations (Categories 2 and 3). While the present findings provide limited evidence of relationships between sham chewing and sow welfare and productivity, further investigation of the relationships between sham chewing and welfare in multiparous sows would assist in appreciating the implications of sham chewing. Given the high level of sham chewing in the present study, research of this nature may be both pertinent from a sow welfare and productivity perspective and prudent in terms of addressing community and NGO criticisms of indoor sow group-housing.

## Neurophysiological correlates of different forms of stereotypic behaviour in a model carnivore

*María Díez-León[1,2], Lindsey Kitchenham[1], Robert Duprey[3], Craig Bailey[1], Elena Choleris[1], Mark Lewis[3] and Georgia Mason[1]*
[1]*University of Guelph, 50 Stone Rd., N1G 2W1 Guelph, Canada,* [2]*Royal Veterinary College, University of London, Hawkshead Lane, AL9 7TA Hatfield, United Kingdom,* [3]*University of Florida, 1149 Newell Dr., Gainesville FL 32611, USA; mdiezleon@rvc.ac.uk*

Stereotypic behaviours (SBs) are common in many carnivore species kept in zoos and conservation breeding centres. SBs are a welfare concern, since typically associated with poor housing. Additionally, SBs are perceived negatively by the public, and can also compromise breeding success. Impoverished housing is known to have deleterious effects on brain development. This has led to hypotheses that SBs reflect housing-related changes in how the brain controls behaviour. However, data on the neural bases of cage-induced SBs to date only come from a few, non-carnivore species. For example, in impoverished-housed deer mice, SBs have been linked to disinhibition in basal ganglia (BG) regions implicated in motor control (e.g. subthalamic nucleus, STN) while individual differences in SBs in horses and C57Bl/6 mice have instead been linked with activation in the nucleus accumbens (NAc): a limbic BG region implicated in compulsive behaviour. We investigated whether similar changes in the BG underlie SBs in a model carnivore, American mink. We raised 16 males in non-enriched (NE) and 16 in enriched (E) cages, and assessed two forms of SB after 2 years: carnivore-typical locomotor-and-whole-body ('loco') SBs such as route-tracing; and an idiosyncratic scrabbling ('scr') with the forepaws. Post-mortem, neuronal activity was analysed via cytochrome oxidase (CO) staining of dorsal striatum (caudate; putamen), globus pallidus (externus, GPe; internus, GPi), STN and NAc). The GPe:GPi ratio was also calculated to assess relative activation of inhibitory/excitatory pathways. NE mink performed more loco SBs ($F_{1,29}$=13.866, $P$=0.0008) but similar levels of scr compared to E mink. NE mink had lower GPe:GPi ratios ($F_{1,24}$=4.723, $P$=0.003) indicting relatively less activity in the indirect pathway, but no one single BG area was significantly affected by housing. Independent of housing effects, elevated CO staining in the NAc predicted higher levels of loco SBs ($F_{1,23}$=6.132; $P$=0.021). Scrabbling, perhaps because it negatively covaries with loco SBs, negatively covaried with NAc activity, but only in NE individuals ($F_{1,12}$=5.013; $P$=0.045). Results, therefore, implicate the limbic system in mink SBs. But they also show that BG function alone cannot fully explain the strong housing effects on loco SBs. Areas outside of the BG must play a role, and these (e.g. prefrontal cortex, hippocampus, cerebellum) should be explored in future work. These results thus highlight the importance of considering different forms of SB separately, and of not relying on individual differences alone when making inferences about the neurological bases of SB.

## Boredom-like states in mink are rapidly reduced by enrichment, and are unrelated to stereotypic behaviour

*Andrea Polanco, Rebecca Meagher and Georgia Mason*
*University of Guelph, 50 Stone Rd E, Guelph, ON N1G 2W1, Canada; apolanco@uoguelph.ca*

Captive animals often live in barren housing that may trigger stereotypic behaviour (SB). Barren housing is also suggested to cause "boredom": a negative state triggered by monotony that increases motivations to obtain stimulation. Farmed mink are excellent models for testing the hypotheses that: (1) monotony causes boredom (as it does in humans); and (2) boredom causes SB (since SB is often reduced by diverse enrichments, suggesting that SB could reflect motivations to obtain stimulation). We previously did so in two independent populations, by exposing mink to diverse stimuli (ranging from rewarding to aversive) when housed in monotonous (non-enriched [NE]) or stimulating (enriched [E]) conditions. We inferred minks' interest in these stimuli from latencies to contact, and time spent oriented towards and in contact with them. As predicted if NE housing causes boredom, interest in all stimuli was significantly higher in NE than E mink (using orientation and contact values). However, SB did not positively predict interest in these stimuli (the first study finding the opposite for contact duration; and the second, no link at all). Using a new, third sample of young adult males (n=32) from an experimental farm, we re-tested this hypothesis, and tested a new hypothesis: that stimulating environments rapidly reverse boredom. Minks' interest in stimuli was tested after seven months in typical commercial, minimally-enriched (ME) cages (containing 1 wiffle ball and 1 shelf, to meet Canadian Codes of Practice), and two days after half the mink were moved to extra-enriched (EE) housing (larger cages with more physical enrichments and shelves). Mixed models tested for housing effects on boredom (mink ID set as a random effect), while Spearman correlations tested whether SB and boredom covaried. Compared to ME mink, contact times were increased ($F$=15.70, $P$=0.0004) and orientation times tended to ($F$=3.03, $P$=0.09), although latencies were not, in EE mink. SB was reduced in EE mink too ($Z$=2.78, $P$=0.006). But again, SB did not predict more interest in stimuli in ME conditions (save for a negative trend with latency to contact: $rho$=-0.32, $P$=0.08); and after the move to EE housing, nor did reductions in interest in the stimuli covary with reductions in SB. Finally, given the contradictory patterns across studies, we pooled data from all three to re-analyse, with greater power, links between SB and boredom-like elevated exploration (in GLMs with 'study' as a blocking factor, and SB and housing as predictors). Overall, there were no significant associations between interest in stimuli and SB. Thus, boredom-like states (i.e. generalised motivations for stimulation) do not seem to cause SB in mink. Our new study also supports the hypothesis that commercial cages induce boredom-like states which can be rapidly reduced by enrichment.

## A shy temperament correlates with respiratory instability during anaesthesia in laboratory-housed New Zealand white rabbits

*Caroline Krall and Eric Hutchinson*
*Johns Hopkins Medicine, Molecular and Comparative Pathobiology, 720 Rutland Avenue, 21205 Baltimore, MD, USA; ckrall2@jhmi.edu*

Stress is well known to negatively affect cardiovascular parameters during anaesthesia. This is especially concerning for prey species, such as rabbits, which are inherently predisposed to nervous temperaments. Whereas anaesthesic precautions can be taken if temperaments are known in advance, as is typical for client-owned rabbits, this benefit is lost in the research setting where not all investigators are familiar with their subjects. This increases the risk of life-threatening complications during experimental surgery. To address this limitation, we sought to determine whether temperament testing (along a shy-bold continuum) impacted quality of anaesthesia. We examined 7 adult, singly-housed, female New Zealand white rabbits who were not previously handled and were scheduled for a brief ophthalmic surgery. Temperaments were assessed using three tools: cage-side behavioural observations, human intruder test (HIT), and novel object test (NOT) (which employed an unfamiliar canine toy tested in advance for interestingness by non-subject rabbits). Subsequently, rabbits were induced with ketamine (30 mg/kg)-dexmedetomidine (40 mcg/kg) i/m, intubated, and maintained on 0.8-1.5% isoflurane. In addition to standard anaesthesic monitoring parameters (i.e. heart rate [HR], respiratory rate [RR], end tidal carbon dioxide concentration, and haemoglobin oxygen saturation), we recorded the latency to loss of righting and pedal reflexes, ease of intubation, and frequency of pre- and peri-operative apnoeic events (i.e. breath holding) every 5 minutes. The first 15 minutes of each anaesthetic period were analysed to eliminate variability between individual surgery durations. We found that 2 rabbits possessed a shy temperament and 5 exhibited boldness, as determined by the results of the behavioural observations and HIT. All rabbits expressed an equal latency to explore the object (Av=16.7 seconds) in the NOT. Shy rabbits exhibited significantly greater variability in RR (Kruskal-Wallis, $H=30.65$, $P=0.00$) and HR ($H=38.86$, $P=0.01$), and more perioperative apnoeic events ($H=5.0$, $P=0.03$) than did the bold group. Thus, in our small pilot study, the temperament test differentiated a stress-prone subpopulation of rabbits, which correlated with an increased rate of apnoea and decreased cardiovascular stability during anaesthesia. Our prevalence of 28% (2/7) shy rabbits was similar to published rates of generalised anxiety disorders in humans (15-30%). NOT was likely ineffective as the standard housing of rabbits is rather barren, so the excitement of new enrichment may have outweighed any underlying anxiety toward the novelty. Future research will apply this method to a larger population of rabbits, and assess if the addition of pre-anaesthetic anxiolytics will ablate the cardiovascular consequences of stress.

## Are towel wraps a better alternative to full-body restraint when handling cats?

*Carly Moody, Cate Dewey and Lee Niel*
*Ontario Veterinary College, University of Guelph, Department of Population Medicine, Stewart Building, 43 McGilvray Street, N1G 2W1, Guelph ON, Canada; cmoody@uoguelph.ca*

Owned, shelter, and laboratory cats are commonly handled and restrained for routine health examinations and necessary procedures. Some handling techniques can increase fear and aggression, resulting in inadequate physical examinations, which is a health and welfare concern. Towel wraps are often recommended as alternatives to high levels of restraint, based on the assumption that they are a low-stress cat handling method. This study used cats from 3 shelters to examine cat responses to restraint, using the following three treatment groups: (1) full-body (side placement with head and legs immobilized by handler; known negative; n=17); (2) head-in towel wrap (n=19); and (3) head-out towel wrap (n=18). All cats were also held with passive restraint (upright with little restraint allowing movement of head, body and limbs; control) for comparison, with restraint order counterbalanced across cats. Cats had been in the shelter for >3 days prior to the study. During restraint, behavioural (lip licks/minute, vocalizations, ear position) and physiological (respiratory rate, pupil dilation ratio) parameters were examined for all cats, except lip licks/min, pupil dilation, and ear position for head-in towel wrap cats. Before testing, all cats were first categorized as friendly or unfriendly based on interactions with a stranger. Treatments were compared using generalized linear mixed models (lip licks, vocalizations), and linear mixed models (respiratory rate, pupil dilation ratio) that included sex, age and friendliness as covariates, and shelter as a random effect. A Cochran-Mantel-Haenszel $\chi^2$ test was used to analyze ear position. During the first 15 seconds of handling, full-body restrained cats showed more side and back ear positions in comparison to passive (*P*=0.043), and head-out towel wrap (*P*=0.034) restrained cats; differences between passive and head-out towel wrap were not detected. Respiratory rate showed a treatment by friendliness interaction (*P*=0.045). On average (95% CI), friendly full-body restrained cats had a respiratory rate that was 11.1 (2.0, 20.3; *P*=0.018), and 10.0 (1.0, 19.0; *P*=0.03) breaths per minute greater than friendly cats that were restrained with head-in and head-out towel wraps, respectively. Average respiratory rates for friendly head-in towel wrap restrained cats were 7.6 (0.8, 14.4; *P*=0.03) and 9.8 (0.6, 18.9; *P*=0.037) breaths per minute less than unfriendly passive, and unfriendly head-out towel restrained cats, respectively. No other significant treatment effects were found. The results suggest that a head-in towel wrap may provide a good alternative for friendly cats, and a head-out towel wrap may provide a good alternative for both friendly and unfriendly cats, since they resulted in fewer negative responses than full-body restraint, and responses were comparable to passively restrained cats.

## Effect of owner presence on dog responses to a routine physical exam in a veterinary setting

*Anastasia Stellato[1], Tina Widowski[2], Cate Dewey[1] and Lee Niel[1]*
*[1]University of Guelph, Department of Population Medicine, 50 Stone Road E, Guelph, Ontario, N1G 2W1, Canada, [2]University of Guelph, Animal Biosciences, 50 Stone Road E, Guelph, Ontario, N1G 2W1, Canada; astellat@uoguelph.ca*

Many companion dogs show signs of fear during veterinary visits. Dogs are often removed from their owners for aspects of exams or procedures, but the effect of this removal has not previously been examined and anecdotal reports suggest that it can be both beneficial and detrimental. Therefore, we assessed the effect of owner presence on behavioral and physiological indicators of fear during a routine physical exam in a companion animal veterinary clinic. Owned dogs and their owners were recruited locally, ages ranged from 1-8 years old, 19 were male and 13 were female. Sex and age were balanced across treatments and the dogs were randomly allocated to receive a standardized physical examination with (n=16) or without (n=16) their owner present. Dog behavior was recorded and responses were assessed by a blinded observer during each exam phase (head exam, lymph node palpation, abdominal palpation, temperature, heart rate, and respiratory rate). For behavioral measures, mixed poisson and logistic regression models assessed the effects of owner presence, exam phase, sex, and age, with dog as a random effect. For physiological measures, linear regression models were used. When owners were present, dogs had a lower frequency of certain fear indicators. In particular when the owner was present vs absent, they had a reduced rate [95% CI] of vocalizations (e.g. whining, barking) (present, 0.11 vocalizations/min [0.03-0.39] vs absent, 1.84 vocalizations/min [1.2-2.8]; $F_{(1,29)}=19.84$, $P<0.0001$), reduced mean temperature (present, 37.2 °C [36.7-37.7] vs absent, 38 °C [37.5-38.5]; $F_{(1,27)}=6.13$, $P<0.0198$), and reduced heart rate (present, 82.4 bpm [74.2-90.6] vs absent, 98.6 bpm [89.7-107.5]; $F_{(1,27)}=7.58$, $P<0.0104$). Rate of lip licking showed an interaction with age ($F_{(1,28)}=6.39$, $P<0.0174$), and was lower with the owner present in older dogs. Interestingly, dogs yawned more when their owner was present (0.24 yawns/min [0.08-0.69] vs absent (0.05 yawns/min [0.01-0.21]; $F_{(1,30)}=4.25$, $P<0.048$). Owner presence did not significantly influence posture reductions, avoidance, escape attempts, or respiratory rate. However, exam phase also influenced fear indicators, including the frequency of posture reductions ($F_{(5,150)}=3.49$, $P<0.0052$), avoidance ($F_{(4,152)}=8.71$, $P<0.0001$), escape attempts ($F_{(5,154)}=5.68$, $P<0.0001$), and lip licking ($F_{(5,155)}=5.7$, $P<0.0001$). Overall, fear indicators appeared to be higher during the head exam, lymph node palpation, and body palpation phases of the exam in comparison to assessments of temperature, heart rate, and respiratory rate. These results suggest that fear levels are reduced in dogs when their owners are present during an examination, and that owners should be encouraged to remain with their dog whenever possible during veterinary visits.

## Effect of background noise on cat responses to a routine physical exam in a veterinary setting

*Nicole Furgala[1], Carly Moody[1], Hannah Flint[1], Shannon Gowland[2] and Lee Niel[1]*
*[1]University of Guelph, Population Medicine, 50 Stone Road East, N1G 2W1, Guelph, ON, Canada,*
*[2]University of Guelph, Hill's Pet Nutrition Primary Healthcare Centre, 50 Stone Road East, N1G 2W1, Guelph, ON, Canada; niell@uoguelph.ca*

In previous research, cats have been found to show signs of fear and aggression during veterinary appointments, and unfamiliar and startling stimuli, such as noise, might contribute to negative responses. We investigated whether common veterinary background noises influence cat responses to a routine physical examination. Owned cats received a physical examination with (n=16) or without (n=16; control) exposure to a pre-recorded noise track. The 9-min track included sounds of dogs barking, kennel doors shutting, and people talking, and had a maximum sound level of 71 dB, which is similar to previous recordings from veterinary clinics. Testing was conducted within a standard examination room at a veterinary clinic. Before the examination, all cats were categorized as friendly or unfriendly based on interactions with a stranger. Behavioural responses (lip licks, side and back ear positions, crouched body posture, struggling, escape behaviour) were recorded during the five consecutive examination phases (weighing, head examination, temperature assessment, heart rate assessment, and respiratory rate assessment combined with body examination), and later scored by a blinded observer. Heart rate and respiratory rate were assessed once during their related examination phase, and photos of the eyes were captured during the head exam to measure pupil diameter. Mixed linear, logistic and Poisson regressions were used to compare responses between treatments, and where appropriate, models included covariates of sex, age, and friendliness, with cat ID as a random effect. When compared to control cats, cats exposed to the noise track had higher heart rates (+36.3 bpm; 95% CI: 10.3, 63.4; $P<0.01$) and respiratory rates (+16.3 bpm; 95% CI: 6.2, 26.3; $P<0.005$). Furthermore, there was an interaction between age and respiratory rate, with older, unfriendly cats having higher respiratory rates than younger friendly cats ($P<0.0005$). No behavioural differences were identified between noise and control treatments. However, cats showed an increased rate of lip licks and escape attempts during some exam phases. For example, during weighing, rates of lip licking and escape attempts were 2.1 (95% CI: 1.1, 3.8; $P<0.05$) and 7.0 (95% CI: 3.3, 14.7; $P<0.001$) times higher, respectively, than during the phase of the exam when heart rate assessment took place. Therefore, ceiling effects do not appear to have been an issue in terms of detecting differences in behavioural responses to noise treatment. These results suggest that background noise has a negative effect on cats during veterinary physical exams, but that aspects of the exam itself have a greater influence. Since examination order was standardized in the current study, further research is needed to explore whether differences between phases were related to the activities occurring during that phase of the examination, or due to habituation to handling over the course of the examination.

## The ubiquity and utility of camels

*Michael Appleby*
*University of Edinburgh, Jeanne Marchig International Centre for Animal Welfare Education,*
*Easter Bush, EH25 9RG, United Kingdom; michael.appleby@ed.ac.uk*

Camels have been contradictorily characterised as either irreplaceably useful ('The ship of the desert') or comically clumsy ('A horse designed by a committee'). This talk will illustrate their wide value in phrase and fable. Some examples demonstrate an everyday familiarity with camel behaviour ('The camel's nose' and 'Better to have the camel inside the tent'). Others demonstrate familiarity with camels, but are rhetorically concerned with human behaviour ('Easier for a camel to go through the eye of a needle' and 'Straining at a gnat but swallowing a camel'). Yet others are superficially about animal management, but really offer sociological understanding with camels as a vehicle ('The last straw that breaks the camel's back', 'The men who say that their camel is slowest' and 'The father who left his camels to his children'). All speak of a world in which humans and animals have longstanding, close and beneficial interactions, a world that is variable in its survival and expression today.

# Authors index

Printed in the United States
by Baker & Taylor Publisher Services